Advances in Legume Research: Physiological Responses and Genetic Improvement for Stress Resistance

(*Volume 1*)

Edited by

Phetole Mangena

Department of Biodiversity
Faculty of Science and Agriculture
University of Limpopo, Limpopo Province
Republic of South Africa

Advances in Legume Research: *Physiological Responses and Genetic Improvement for Stress Resistance*

Volume # 1

Editor: Phetole Mangena

ISSN (Online): 2737-4890

ISSN (Print): 2737-4882

ISBN (Online): 978-981-14-7962-5

ISBN (Print): 978-981-14-7960-1

ISBN (Paperback): 978-981-14-7961-8

need for a court order if at any point you breach any terms of this License Agreement. In no event will any delay or failure by Bentham Science Publishers in enforcing your compliance with this License Agreement constitute a waiver of any of its rights.

3. You acknowledge that you have read this License Agreement, and agree to be bound by its terms and conditions. To the extent that any other terms and conditions presented on any website of Bentham Science Publishers conflict with, or are inconsistent with, the terms and conditions set out in this License Agreement, you acknowledge that the terms and conditions set out in this License Agreement shall prevail.

Bentham Science Publishers Pte. Ltd.
80 Robinson Road #02-00
Singapore 068898
Singapore
Email: subscriptions@benthamscience.net

BENTHAM SCIENCE

CONTENTS

FOREWORD

Legumes form a major component of daily meals all over the world , especially in the developing world where in some cases, they are the stable food. They are also used in animal feeds. As the world population increases, more demand is placed on basic foods like legumes. Global warming has a detrimental effect on agricultural food production. In addition, people have become more concerned about the quality of the food they consume. These necessitate more research in agricultural food production, with legumes being no exception.

This book, "Advances in legume research – physiological responses and genetic improvement for stress tolerance" offers a reference to those who want to improve their legume production, those who seek better legumes to produce and those who seek foods with improved nutritional quality. Research topics covered include usage, cultivation, basic research, as well as genetic transformation in order to induce biotic and abiotic stress tolerance.

Genetic manipulation of food crops is still not fully acceptable by some people and countries. To those who have this view, the book shows that some basic physiological research methods can still be used to improve legume crops.

Phatlane William Mokwala
Department of Biodiversity
University of Limpopo, Limpopo Province
South Africa

PREFACE

Legumes fall within the group of pod producing grain plants, belonging to the family *Fabaceae* . Plant species found inside this taxon are of high significance globally because of their relatively higher amounts of quality proteins, carbohydrates, fibre, and essential oils, that are contained within the seeds. Legume seeds contain less fats and no cholesterol, and they are a potential source of innumerable food/pharmaceutical supplements, feed manufacturing ingredients, biofuel, and are positively related to sustainable agriculture due to their association with nitrogen fixing microorganisms. It is, therefore, due to the above mentioned reasons that; they serve as model crops for functional studies in the trait (growth and yield) improvement of crops and the physiological/genomic development of resistant varieties against climate change-induced stress.

Of all the ways that climate change inflict harm on crops, legumes are among the most vulnerable and highly sensitive groups of oilseed crops worldwide. Abiotic stress and biotic stress-based reductions in the growth and yield of these crops, particularly, soybean, mung bean and cowpea, cause greater negative impacts on food security, health, and the import-export rates in many countries. Although these crops were recently rated just around 767th of the most traded products according to the OEC (Observatory of Economic Complexity, 1995–2017), they all remain a positive driver for sustainable growth and development of many countries' gross domestic product (GDP), especially in the developing African region.

All authors believe that readers will, therefore, receive and appreciate the insights provided by Advances in Legume Research- Physiological Responses and Genetic Improvement for Stress Resistance from different individuals with high expertise, and who are specialists in the area. The authors are of different scientific backgrounds, which is very important for the diversity of views, bringing new ideas and sharing new important original information covering perspectives from various fields of legume physiology and genetics. Legumes research should undoubtedly be continued, particularly because these crops also have a narrow genetic pool, consequently making them highly sensitive to various stress factors.

Numerous genetic modification tools gained popularity in recent years, but, inefficiencies and losses of beneficial genes still persist in many breeding systems. Some approaches are widely criticised, and others appear as viable alternatives, simply because they receive considerable attention from consumers and researchers. There are widespread untested assumptions of genetic instability of genetically modified organisms, and possible carcinogenic effects believed to emanate from genetically modified plants. Therefore, investigations on the evaluation of chemical compositions, growth, development and reproduction of GMOs and non-GM legume crops is perhaps the most fundamental service to mankind and the increasing populations.

ACKNOWLEDGEMENTS

The paramount goal of delivering a comprehensive book that clearly elucidates the understanding of mechanisms involved in plant's genetic and physiological responses to stress was made possible by all authors. As such, we are very much grateful to all the authors and everyone who provided their meaningful contributions. I express my special thanks and appreciation to Dr. Phatlane William Mokwala and Prof. Roumiana Vassileva Nikolova for their continued mentorship and support. We express our special thanks and appreciation to Fariya Zulfiqar, Publication Manager for the support and help in making this goal achievable.

Phetole Mangena
Department of Biodiversity
University of Limpopo, Limpopo Province
South Africa

List of Contributors

Andries Thangwana
Irrigation and Climate Control Department (ICC), Flamingo Horticulture, Plot 25 Delarey Farm, Syferbult Road, Tarlton 1749, South Africa

Arinao Mukatuni
Department of Biodiversity, School of Molecular and Life Sciences, Faculty of Science and Agriculture, University of Limpopo, Private Bag X1106, Sovenga 0727, South Africa

Erlafrida Ramokgopa
Department of Biodiversity, School of Molecular and Life Sciences, Faculty of Science and Agriculture, University of Limpopo, Private Bag X1106, Sovenga 0727, South Africa

Esmerald Khomotso Michel Sehaole
Department of Biodiversity, School of Molecular and Life Sciences, Faculty of Science and Agriculture, University of Limpopo, Private Bag X1106, Sovenga 0727, South Africa

Lifted Olusola
Department of Biodiversity, School of Molecular and Life Sciences, Faculty of Science and Agriculture, University of Limpopo, Private Bag X1106, Sovenga 0727, South Africa

Paseka Tritieth Mabulwana
Department of Biodiversity, School of Molecular and Life Sciences, Faculty of Science and Agriculture, University of Limpopo, Private Bag X1106, Sovenga 0727, South Africa

Phatlane William Mokwala
Department of Biodiversity, School of Molecular and Life Sciences, Faculty of Science and Agriculture, University of Limpopo, Private Bag X1106, Sovenga 0727, South Africa

Phetole Mangena
Department of Biodiversity, Faculty of Science and Agriculture, University of Limpopo, Limpopo Province, Republic of South Africa

Nwagu Rodney Mashamba
Agricultural Research Council, Agronomy and Technology Transfer Department, Private Bag X82075, Rustenburg 0300, Republic of South Africa

Phumzile Mkhize
Department of Biochemistry, Microbiology and Biotechnology, School of Molecular and Life Sciences, Faculty of Science and Agriculture, University of Limpopo, Sovenga 0727, South Africa

Samuel Tebogo Posie Peta
DSI-NRF Centre of Excellence for Invasion Biology Department of Botany and Zoology, Faculty of Science, Stellenbosch University, Private Bag X1, Matieland 7602, South Africa

CHAPTER 1

Breeding of Legumes for Stress Resistance

Phetole Mangena[*]

Department of Biodiversity, School of Molecular and Life Sciences, Faculty of Science and Agriculture, University of Limpopo, Private Bag X1106, Sovenga 0727, South Africa

Abstract: As grain legumes continue to be used for various food and health forms, after suitable processing and manufacturing of legume-based products, aspects such as growth, yields, physiological stress and genetic manipulation remain significant topics for the enhancement of their utilisation, to explore new potential and diversify their genetic resources. Future research focusing on the physiological response and genetic improvements of legumes need to be prioritised to improve the utilisation and nutritional quality. The purpose of this chapter is to serve as an introduction to advances made in grain legumes, that are presented in various chapters of this book. The discussion is generalised and intended to provide a comprehensive view on the effect of stress on legume growth and yields. Included in this chapter are (a) a brief discussion on legume origin and classification, (b) brief survey on legume growth, yield and the impact of stress (biotic or abiotic stress) and (c) overview on breeding strategies available for genetic improvement of grain legume species, both conventional and non-conventional technologies.

Keywords: Abiotic stress, Biotechnology, Biotic stress, Breeding, *Fabaceae*, Growth, Legumes, Yield.

THE LEGUME (*FABACEAE*)

The Fabaceae family contains over 650 genera and 20,000 species. This plant family is of greatest importance to world agriculture after the *Poaceae* family. In this volume, I will focus on this *Fabaceae* family, but only pay a special emphasis on plant species in this family that are used as edible bean seeds (Fig. **1**). These selected bean species are used as food crops, directly or indirectly in the form of ripe-mature or unripe-immature pods, as well as mature and immature dry seeds. Most cultivated grain legumes belong to the two natural tribes; the *Vicieae* and *Phaseoleae*, both consisting of species and exhibiting phylogenetic characters as indicated in Table **1**.

[*] **Corresponding author Phetole Mangena**: Department of Biodiversity, School of Molecular and Life Sciences, Faculty of Science and Agriculture, University of Limpopo, Private Bag X1106, Sovenga 0727, South Africa; Tel: +2715 268 4715; Fax: +2715 268 4323; E-mails: phetole.mangena@gmail.com & Phetole.Mangena@ul.ac.za

Phetole Mangena (Ed.)

Both the *Vicieae* and *Phaseoleae* species have a combination of hypogeal/epigeal germination system and the herbaceous plant habit [1].

Biological characters such as those highlighted above, clearly indicate a simple inherent genetic control, additionally signifying the fact that many species within the tribes are restricted to one system or may interchange between epigeal and hypogeal germination systems among represented species. These and other key diagnostic characters (Table **1**) are a representation of residual traits evolved from ancestral associations [2, 3]. The fruits, which are of pod type vary from dehiscent to indehiscent with a morphological diversity, which is translated into notable variations in seed dispersal mechanisms, such as ornithochory, hydrochory, autochory, anemochory *etc* [4]. Much of the diversity is exploited in agriculture, especially for the nine (9) annual grain species widely cultivated for commercial or domestic purposes, that include the dry bean, common bean, pea, lentil, mung bean, faba bean, cowpea, pigeon pea, and soybean. All these crop species have grain quality that is suitable for industrial processing.

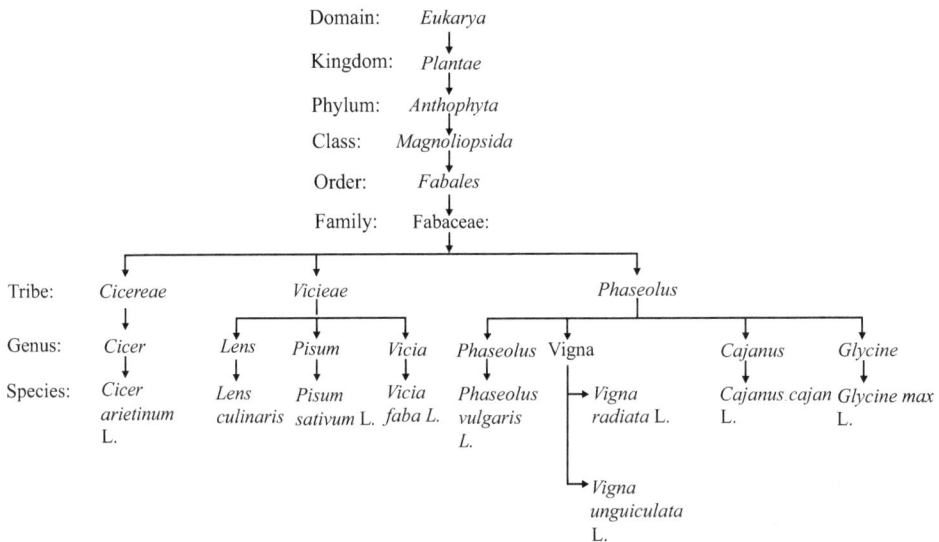

Domain:	*Eukarya*
Kingdom:	*Plantae*
Phylum:	*Anthophyta*
Class:	*Magnoliopsida*
Order:	*Fabales*
Family:	Fabaceae:

Tribe:	*Cicereae*		*Vicieae*			*Phaseolus*	
Genus:	*Cicer*	*Lens*	*Pisum*	*Vicia*	*Phaseolus* *Vigna*	*Cajanus*	*Glycine*
Species:	*Cicer arietinum* L.	*Lens culinaris*	*Pisum sativum* L.	*Vicia faba* L.	*Phaseolus vulgaris* L. → *Vigna radiata* L. → *Vigna unguiculata* L.	*Cajanus cajan* L.	*Glycine max* L.

Fig. (1). A summary of taxonomic classification scheme of selected grain crops and their botanical names (genus and species).

DOMESTICATION OF GRAIN LEGUMES

The crop species represented above form part of what is now known the civilisation and initiated human dominion over natural plant genetic resources on earth. The gathering and domestication of desirable wild plant species began over 10,000 years ago, leading to well-coordinated breeding practices where species were selected and propagated for greater and more convenient food, as well as medicinal supply. According to phylogenetic evidence-based descriptions provided by Schrire *et al.* [5], Lopez *et al.* [4] and Moteetee and Van Wyk [6] in legumes, the selection of crop species for domestication was based on ancestral associations, their usefulness in the primitive economy and the ease of domestication, especially on the simplicity in which selected species could be propagated. The majority of the earliest domesticated species could be identified by their native and endemic traditional uses in certain areas. The distribution and utilisation of cultivated crop species differ according to the age of domestication, period enabling novel introduction to other areas for similar purpose and areas of wider distribution and occurrence of the wild ancestral populations.

Hartmann *et al.* [7], reported that peas and lentils were the earliest domesticated legume crops, together with wheat and barley cereals in the eastern part of the world. In the far east, millet appears to be the first domesticated crop followed by rice, meanwhile squash and avocado were the first domesticated crops in the central and southern parts of America. These plants are followed by corn, bean, pepper, tomato and potato in the same region, which now serve as some of the major commodities and widely cultivated crops worldwide. Domestication and spread of legume crops across the world did not only enable sustainable food and medicine supply, but also permitted effective capture and recycling of energy from the sunlight [8].

Raven [8] indicated that these may include the capture of CO_2 (for incorporation into carbon skeletons used for carbohydrate synthesis and synthesis of other carbon-containing primary and secondary metabolites with a carbon backbone) and improvement in soil fertility through nutrient recycling, when legumes die and replenish the mineral nutrients back in the soil. The spatial distribution of these legume crops still influences the structure and functioning of other plant populations and communities, particularly due to a mutualistic symbiotic relationship with nitrogen fixing bacteria.

Table 1. Botanical names and common names of some of the grain legumes and their tribes, found under the family Fabaceae.

Tribe	Species	English Common Name	Tribe Phylogenic/ Taxonomic Characters
Vicieae/ Fabeae	*Lens culinaris*	Lentil	• Solitary or racemose, pubescent flowers, • Paripinnate leaves, with tendrils, • Well-developed stipules, • Usually low or climbing herbs [1].
	Pisum sativum L.	Pea	
	Vicia faba L.	Broad bean/ faba bean	
	Vicia ervilia L.	Bitter vetch	

(Table 1) cont.....

Cicereae	*Cicer arietinum* L.	Chickpea	• Pink flowers, with anthocyanin pigmented stems, • White or beige-coloured seeds with ram's head shape [2].
Dalbergieae	*Arachis hypogeae* L.	Peanut	• Plestomorphic flowers (free keel petals, staminal filaments partly fused, • Filaments without basal fenestrae, • Inflorescence of determinate growth [3].
Genisteae	*Lupinus luteus* L	Yellow lupin	• Hilum positioned laterally with straight hilar groove, • Pods oblong or linear-oblong, with transversely ovate seeds, • Small ovaries producing larger pods [4].
	Lupinus albus L.	White lupin	
Indigofereae	*Cyamopsis tetragonoloba* L.	Guar	• Presence of re-carmine flowers, lacking anther hairs, • Perforated pollen, • Paniculate inflorescences [5].
Phaseoleae	*Canavalia gladiate*	Sword bean	• Style tip is expanded and spoon, • Bearded style, • Standard petals with appendages, • Hilum covered with white spongy tissues [6].
	Canavalia ensiformis L.	Jack bean/ horse bean	
	Cajanus cajan L.	Pigeon pea	
	Glycine max L.	Soybean/ soja	
	Mucuna pruriens L.	Velvet bean	
	Phaseolus acutifolius	Tepary bean	
	Phaseolus lanatus L.	Butter bean	
	Phaseolus vulgaris L.	Dry bean	
	Vigna radiata L.	Mung bean	
	Vigna aconitifolia	Moth bean	
	Vigna mungo L.	Black gram	
	Vigna umbellate	Rice bean	
	Vigna unguiculata L.	Cowpea	

GROWTH, YIELD AND STRESS

The growth and development of legume crops from a small grain seed to a mature plant requires a precise and highly organised succession of cellular, genetic, physiological and morphological events. Starting as a single fertilised gamete, plant cells divide, grow and differentiate into an astonishingly complex miniature plant called an embryo, packaged within a seed. In the end, the seed will germinate and give rise to the complex organisation of seedling tissues and organs that divide and grow into a mature plant that later flowers, bears fruits, disperses the seeds, senesces and eventually dies. According to Raven [8], all these events, including the biochemically and environmentally modulated processes constitute plant development and growth. Understanding these processes is one of the major goals of crop physiology. The pattern of changes experienced by cells, tissues and organs is more genetically controlled. Like other plants, legumes also experience a well-coordinated growth of tissues, which is subjected to control at various distinct levels. Such control levels include intrinsic control operating at both intracellular and intercellular level (*e.g.* gene or protein expression and hormonal regulations), and extrinsic extracellular controls outside the organism, functioning to convey information about the environment [9].

Legumes likewise often encounter unusual or extreme environmental conditions like any other plants. Crops in the northern latitudes for example, experience extreme low temperature, while those in the tropical Savannas may experience scorching temperatures, and high levels of harmful UV radiations. These effects are much felt by farmers of agricultural crops, whose plants may experience a period of extended drought (thus leading to disease outbreak and uncontrollable spreads) or their roots subjected to high salt concentrations in the soil. Unfortunately, plants are rooted in the soil and consequently they cannot escape adverse environmental conditions in their vicinity. Rather, they use various stress response strategies to adapt, survive and grow under these hostile conditions [10]. A major problem is that both natural and anthropogenic activities continue to add and exacerbate the number of stress factors that plants must cope with in their environment (Fig. **2**). Dakhovskis *et al.* [11] reported the physiological adaptation of cultivated plants following exposure to naturally and anthropogenically induced environmental stress. The stress factors affected individual biological characteristics of the plants, emphasising a strong impact on plant homeostatic mechanisms and weakening of plant response to the induced stress.

Fuchs *et al.* [12] also analysed genome damage, infertility and meiotic abnormalities caused by agricultural expansion and increased utilisation of agrochemicals, releasing heavy metals into the environment, pathogen spread and contamination by pharmaceutical or industrial residues. Thus, all cropping systems need to elaborate on the system's productivity and sustainability in addition to profitability. All stakeholders should be concerned about conserving the quality of the environment and maintaining soil fertility as much as they pay attention on the quality and quantity of yields. The yield potential of many grains is seldom achieved due to unsuitable cultivated species and inadequate crop management to cope with stresses [13]. The development of fruit pods and seeds is strictly genetically and physiologically modulated. Therefore, the exposure of flowering plants to stress usually cause ovaries and embryo development to abort, or slowed down by hormones that are responsible for the coordination of normal seed and fruit development. The formation of fruits and seeds or the overall yields in grain legumes is linked with irreversible anatomical changes and an aging process. Physiological effects associated with these major changes may include total dry matter, leaf area, photosynthetic rate, stomatal conductance, respiration rates, internal CO_2 metabolism and leaf water potential, all, which have negative impacts on yields [14].

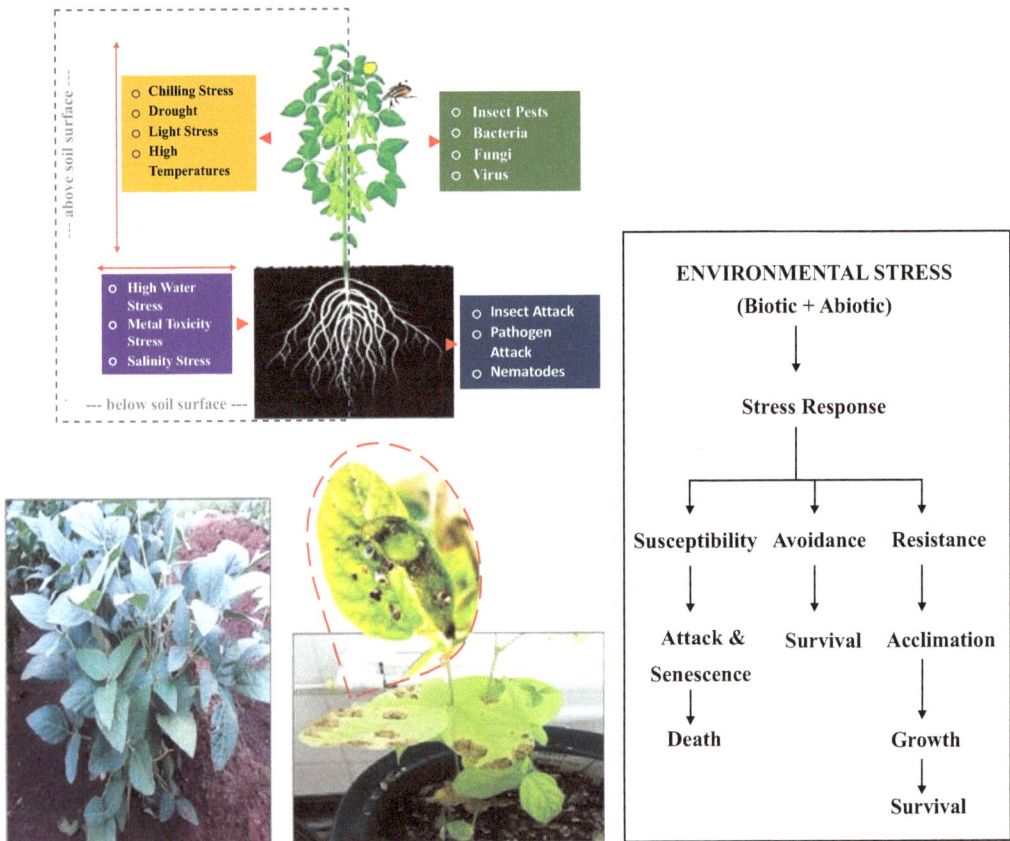

Fig. (2). An overview of the relationship between environmental stress (biotic and abiotic stress) threating plant survival, growth or yield.

BREEDING OPPORTUNITIES FOR SPECIFIC ADAPTATIONS

Grain legumes continue to occupy a crucial position in legume-based diet of many population's nutrition, health and welfare, mainly as a source of proteins, minerals, carbohydrates and vitamins. It has been confirmed that the consumption of grains contributes to a balanced diet and can prevent the progression of cancer and other chronic diseases. The intensification of legume agriculture has also led to major characteristic changes in the agroecological systems [1, 14], enhancing pathogen generation as well as the spread of many biotic and abiotic stress factors. Abiotic and biotic stress agents adversely affect crop growth, cause rapid depletion of natural genetic resources, cause reduction in arable land and are the basis for accumulation of pollutants in the environment. Human population are still yet to face challenges on mitigating the damage caused by myriad of anthropogenic activities, including those caused by unsustainable agricultural practices. These factors already cause major negative impacts on the natural environment and to human/animal health [15].

These effects are, furthermore, exacerbated by the consequences of climate change. The only apprehension for breeders and scientist is that, positive identification and selection of superior genetic resources showing resistance to these stress constraints are required. So far various traditional and modern methods have been used to recover plants that have unique and desirable genetic properties, for example, plants modified through genetic transformation. Researchers worldwide have to continue optimising breeding techniques to expand the number of legume species amenable to genetic improvement.

BREEDING OF LEGUMES FOR ABIOTIC STRESS

In 2006, the United Nations conference on climate change predicted extended drought seasons in most parts of Africa, due to climate change. Furthermore, it was highlighted that agricultural production will suffer more as a consequence of this frequent climate fluctuations [16]. However, since their initial domestication, more legume crops have been subjected to intensive selections and breeding of varieties that contain crucial agronomic traits. These included a set of characters that made them adaptive to adverse environmental conditions and enhanced their growth, quality and quantity of yields. Such improved plant characters included changes in apical dominance, production of enlarged sizes and numbers of roots, stems, leaves, fruits and seeds. But, it is common knowledge that abiotic stress is responsible for major growth and yield losses in many legume crops. Amongst these, drought has been recorded as the most damaging kind of abiotic stress and most crop plants are highly susceptible and sensitive to drought than any kind of abiotic constraint condition [8, 9, 17].

Herbaceous plant crops such as soybean and cowpea are among the grains that easily get injured by moderate or brief exposure to drought stress, immediately exhibiting one or more metabolic dysfunctions. Freitas *et al.* [14], reported reduction on several growth parameters of cowpea following moderate and severe water restrictions. In addition, Wijewardana *et al.* [18] evaluated whether the effects of water deficit stress on parental soybean plants may be transmitted to the F_1 generation. The results showed that, seed germination and seedling development in F_1 generation were affected by the lasting effects of soil moisture stress that took place originally on affected parent plants. The findings emphasised a key role played by seed weight and storage reserves during germination and seedling growth. Thus, concluding that, optimal water supply during fruiting and seed filling period is beneficial for enhancing seed quality and vigour/ viability characteristics. As predictions continue to estimate that climate change will be responsible for 20% increase in water scarcity due to the occurrence of poorly distributed torrential rains, droughts, and high temperatures. This will severely affect crop development and yield, as already seen in China, India and the United States which all serve as the largest global producers of grain crops [19]. Thus, the breeding of legume crops should include genetic improvement for salinity, heat, light, metal toxicity and chilling stress tolerance.

Breeding for Salinity Stress Tolerance

Soil salinity remain one of the major factors adversely affecting crop yields worldwide. Predictions estimates that, about 1 billion hectares of irrigated land is affected by salinity and the problem is increasing at a rate of about 10% per annum. Legume crops show high sensitivity during vegetative and reproductive stages primarily from the abundance of sodium chlorite (NaCl) from irrigated soil or natural accumulation [20, 21]. Salinity stress cause metabolic dysfunction by causing nutritional imbalances, osmotic stress effects, ion toxicity, decreased photosynthetic rates and cause severe necrosis and chlorosis [21]. From the physiological and genetic aspects, salinity stress is a complex trait, therefore, an integrated approach that use the existing genetic resources, diversity and novel sources to create new varieties is required. Sehrawat *et al.* [22] reported that, continued screening should be frequently adopted to select salt-tolerant germplasm to develop better performing genotypes.

Breeding for Temperature Stress Tolerance (Heat and Chilling Stress)

Plants exhibit a wide range of sensitivities to extreme temperatures. Both chilling stress and heat stress have detrimental effects on plant growth and productivity. Heat stress has deleterious effects on the morphology, physiology and reproductive growth of plants. But, the reproductive phase is the most vulnerable stage during the period of stress, which reduces crop yields. According to Wang *et al.* [23] and Bita and Gerats [24], the impact that heat stress have on plant reproduction include reduced pollen viability/mortality, ovule infertility, flower abortion, impaired fertilisation and reduced seed filling leading to decreased seed sizes and yield losses. These observations confirm reports, that daytime temperature above 35°C caused substantial reductions in anthesis and pod setting leading to complete failure of the reproductive phase [25]. Each legume crop species has its unique set of temperature requirement for growth and development. This includes an optimum low temperature at which the plant grows and performs most efficiently without sustaining chilling stress injuries. Under cold stress, vegetative growth of many legumes get severely affected at temperature ranging between 4 to 15°C. Legumes such as chickpea and pigeon pea show high sensitivity to chilling and frost induced stress. Injuries sustained completely inhibit photosynthesis and cause production of reactive oxygen species (ROS) [26]. Therefore, genotypes surviving and reproducing under this temperature range (< 15°C) are highly desirable and may be selected as potential genetic resources for the development of chilling stress tolerant cultivars.

Fig. (3). Anthracnose symptoms on soybean fruit pods as indicated by Tom Allen, Extension Plant Pathologist. **(A)** & **(B)** Soybean plants showing a possible viral infection. **(C)** Seeds attacked by soybean podworm. **(D)** Example of soybean plants grown in a field without any observable disease symptoms.

BREEDING OF LEGUMES FOR BIOTIC STRESS RESISTANCE

Climate change has increased challenges experienced in agriculture by intensifying the spread of diseases affecting legume and cereal grain crops (Fig. 3). It has been widely reported that biotic stresses occur at different intensities across all cultivated agricultural lands worldwide. For example, Anthracnose infections caused by *Colletotrichum* spp. can infect stems, leaves and pods of soybean causing minimal effects on yields (Fig. **3A**) [27]. The occurrence of diseases caused by bacteria, fungi, viruses and constant crop attacks by insects, weeds and nematodes is also increasing at a very alarming rate. These stress factors apparently cause reductions in the growth and yields of many crops. Therefore, for farmers and consumers to cope with biotic stress, plant breeding programmes have to adopt new strategies to rapidly and efficiently develop new cultivars with resistance.

Breeding for Insect Pests

Grain legumes remain the most important source of food and medicine for the increasing populations, currently estimated to have reached 7.8 billion people on earth. The main objective for plant breeders and researchers should be to develop insect-resistant legume crops, that are well-adapted to a diverse range of climates, soil types, and widely cultivated throughout the temperate, tropical and sub-tropical climates. However, the demand for food has been increasing very rapidly, especially with the increasing spread of crop diseases that lead to more than 20% yield losses almost every year. Biotic stress affect the most vulnerable and highly susceptible widely cultivated legumes, such as cowpea. Cowpea is grown in many parts of the world, including west, south and east Africa, Latin America, United States and south east Asia [28]. Although, crop plants naturally express phytoalexins (collection of isoflavonoids and other secondary metabolites) to ward off insects and disease outbreaks, this defense system already proved inefficient for the current evolved sets of attackers (Fig. **3C**).

Cowpea probably suffer most from insect pests, because all parts of the plant at all stages, from seedling to mature plant ready for harvest get infested by insects. A large number of insects pests, which mainly belong to the phytophagous taxa have the ever evolving and increasing gene pool. Its speciation is somehow believed to be intimately linked to their hosts. The insect taxon include *Coleoptera*, *Diptera*, *Homoptera*, *Heteroptera*, *Hymenoptera*, *Lepidoptera*, *Orthoptera* and *Thysanoptera* which comprise about 76 predominate phytophagous families containing at least 20 species of insects each [29]. Modern agriculture should address the lack of genetically improved cultivars and good crop management systems, in order to increase crop yield under such gigantic gene pool of insect pests. Improved cultivars are necessary to achieve abundant food production, high crop quality and make food prices affordable for the poor masses of starving people, particularly living in developing African countries.

Breeding for Pathogen Resistance

One of the most notorious crop pathogens to have caused over 80% yield losses in Europe (1845) is *Phytophthora infestans*. This etiological agent could cause close to 100% yield losses in agricultural production fields, accompanied by huge direct monetary costs of pest control and lost productions estimated at more than $ 3 billion per year globally [30]. Similar *Phytophthora* species that attack pasture legumes were reported by Irwin *et al.* [31]. According to the report, Australia experienced an annual production loss exceeding $ 200 million due to individual species belonging to this taxa. Another devastating pathogen that also cause more than 80% yield losses each year is *Aphanomycetes euteiches*, of the family *Saprolegniaceae*, order *Saprolegniales*. According to Gaulin *et al.* [32], this pathogen infects the cortical tissues in primary and lateral roots of crop legume seedlings. It forms oospores within the root's cortex tissues that causes yellowing of cells followed by browning on root cells, and subsequently leading to the blackening of seedling's hypocotyl. Crop rotation with faba bean, pea and any species of lupin usually leads to a build-up of soil-borne pathogens with long field persistence [30, 31]. Recently, some studies illustrated that, should resistant cultivars not be developed, a large amount of fungicides with agroeconomic and environmental negative effects will not be reduced. Furthermore, billions of dollars of income losses for legume crop growers will also not be avoided, with no possibilities of business recoveries. As long as the spectrum of biotic stresses that may cause crop yield losses is large and continue to diversify as already demonstrated in *Phytophthora* taxa. This threatens the efforts made in providing food, feed, fibre and bioenergy for the increasing world population [32].

AVAILABLE IMPROVEMENT STRATEGIES AND TECHNIQUES

Legumes constitutes a large number of varieties in which many of them are bred and developed for both subsistent and commercial farming purposes worldwide. These crops are improved for pathogen resistance, drought or salinity stress resistance, competitiveness against weeds, high heritability, additive genetic control and shared better performance of individual lines or populations. According to Fritsche-Neto *et al.* [33] selection or hybridisation is often based on hybrid performance for allogamous species when a trait of low hereditability or when its inheritance is based on non-genetic effects. However, many traits will be quantitatively determined by the interactions of a large number of genes expressed uniquely under different environments. Increased variations provide further opportunity for selection and possible hybridisation to produce new genotypes adapted to specific environmental niches.

A recently established revolution in genetic research has led to the development of the most far-reaching applications across the whole range of applied plant biology, especially in the selection and breeding of crop species to sustain plant growth during biotic and abiotic stress. Biotechnology continues to provide and produce improved varieties mainly through transgenesis and other techniques, which allow the introduction of one or more genes for stress tolerance [33]. Some of these methods have become more efficient and highly rapid for the development of new stress resistant cultivars, especially for major crop commodities such as cotton, soybean, rice, maize, wheat and sorghum. These crops are targets of significance for food, beverage, and health industries, as well the potential production of a clean efficient bioenergy, worldwide. Furthermore, according to Haile *et al.* [34] wheat, corn, rice and soybean serve as very important staple commodities and remain crucial for the fight against global food insecurity.

Transgenics

Long-term significant genetic improvement efforts through genetic engineering have been taking place in many private, parastatal and government research laboratories worldwide. Some of these national and international laboratories have thus far registered and distributed a range of improved legume crop lines around the world, accomplishing a number of stress resistance factors and other crop qualities. Such include disease and pest resistance, improved seed quality, biotic and abiotic stress resistance. According to Chandra and Pental [35], among the many different techniques tested for gene delivery to plant cells, only *Agrobacterium*-mediated genetic transformation and particle bombardment have been extensively employed. These techniques can be utilised for or coupled with transient gene expression studies, functional genomics and even CRISPR-Cas9 genome editing technology. CRISPR-Cas9 in short, refers to clustered regularly interspaced short palindromic repeats and CRISPR-associated protein 9.

Gene editing predominantly induces non-homologous end joining (NHEJ), which generates altered genome by random insertions or deletions, and precise recombination products by homology-directed repairs [36].

Particle bombardment-mediated transformation remains one of the most expensive procedures, but rather highly rapid, efficient and suitable for genetic manipulation of many recalcitrant plant species. This method utilises a gene gun employed to allow penetration of the cell wall by foreign particle coated genetic materials containing the gene of interest to be transferred into the host cells [37]. Different plant tissues or organs targeted include pollen grains, embryos, callus cells, seedling epicotyls and hypocotyls, fruits, flowers, and roots [38]. Ivo *et al.* [39] generated stable transgenic cowpea (*Vigna unguiculata*) plants showing a Mendelian transgene inheritance. The transgenic plants were mutated with *ahas* gene coding for acetohydroxyacid synthase conferring high level resistance to herbicide imazapyr. *Cicer arietinum* (L.) was also genetically transformed with an insecticidal crystal protein gene (*crylAc*) taken from *Bacillus thuringiensis* [40].

Particle gun method was, furthermore, used for the genetic manipulation of pigeon pea and soybean using different explant types. Reported studies evaluated the use of different promoters and stacking of genes for various biotic and abiotic stresses [41].

It has been more than two (2) decades since the introduction of genetically modified plants predominantly established through *in vitro Agrobacterium*-mediated transformation. In soybean, the first successful transformation was reported by Hinchee *et al.* [42], using cotyledonary explants with *Agrobacterium* pTiT37-SE plasmid harbouring *pMON9749* gene for herbicide glyphosate tolerance. The success of this method depended upon several factors, which included tissue culture conditions, *Agrobacterium* strain and selected host plant genotypes aimed at receiving the transgenes. To date, this technique has succeeded in the production of high yielding transgenic cultivars, particularly for corn, chickpea, rice, cowpea, as well as a few soybean genotypes [43 - 45]. According to Patel *et al.* [46] and Mehrotra *et al.* [45], this technique is considered the most economic and highly effective method of genetic modification that has been reported so far. The method holds the potential and promise to efficiently regenerate transgenic plants, especially in recalcitrant legume crops.

Mutagenesis

Mutations are primary sources of spontaneous or induced genetic variability resulting from DNA changes that alter the plant's genome. This approach has been used to identify gene function or cause genetic manipulation in plants by involving the use of chemicals, ionizing radiation or specific DNA sequence insertions [33]. Spontaneous mutations are rare and non-targeted as the chance of occurrence is very minimal, meanwhile, targeted mutations are induced *via* treatment of plant tissues with various mutagenic agents. A variety of legume mutant lines, such as those conferring weed control by herbicide tolerance, and those used as germplasm collection for further breeding in the domestication of species for agriculture are already in use and available. Mutation breeding of agronomically important legumes, usually used as models to study legume crop genetics and genomics has also been reported.

Legumes that include models *Medicago truncatula* and *Lotus japonicus* [47] have been reported. Tadege *et al.* [48] also reported mutant populations used as mutant resources for reverse genetics; namely the *Tnt1*-tagged population of *M. truncatula* and the targeting-induced local lesions in genome (TILLING) population of *L. japonicus*. All these mutant lines supply plant breeders with genomic data and a chance to obtain genetic variants from which improved cultivars with higher yield, better grain quality and stronger resistance to pathogens could be achieved [49]. Mutagenic agent alkylated ethylmethane sulfonate (EMS) has been widely used in many plant species to induce single base pair G/C to A/T substitution in nucleotide sequences [50], accompanied by TILLING. The latter is a reverse genetic tool frequently used to identify heteroduplex mismatches in targeted single-stranded DNA sequences [48].

New genetic characteristics were also reported in cowpea, faba bean and pigeon pea through polyploidisation (*i.e.*, chromosome duplication using colchicine mutagenic agent) [51]. Polyploidisation using a chemical mutagen (*e.g.* colchicine, sodium azide (NaM3), Oryzalin, Ethyl Methane Sulfonate (EMS) *etc.*) in the past served as the most important technique used to improve genetic diversity of plants, conferring inherent built-in resistance to various stress factors by modifying morphological and physiological characteristics [51, 52]. In addition to the development of mutant resources, tools such as transcriptomics, proteomics, metabolomics, and bioinformatics are currently being developed with the attempt to translate attained molecular data for use in gene discovery and trait improvement in legume plants. Some of these tools are already beneficially used in plant physiology, genetic population structure, structural genetics, and a variety of other biological fields to enhance the sustainability of agriculture in the face of the ever-increasing food and energy demand [47].

Marker-Assisted Selection (MAS)

Marker-assisted selection or marker-aided selection (MAS) is an indirect method of selecting interest based traits for the marker linked to the traits. RNA/DNA, biochemical and morphological markers linked to grain productivity, quality, biotic stress resistance, or abiotic stress tolerance are selected to improve the efficiency in plant breeding through a precise transfer of genomic regions of interest [53, 54]. However, genomic-wide selection has been widely tested to efficiently accelerate genetic improvement through direct selection during early juvenile plant stages. Performing predictions and selections, early during plant growth was reported to increase the efficiency of potassium and nitrogen absorption under low soil nutrient availability in tropical maize [33, 54]. In rice, DNA based molecular markers that are closely linked to resistance genes were transferred to single genotypes bestowing to various types of biotic and abiotic stress factors [53]. According to Das *et al.* [53], this method can be used to rapidly accelerate the advancement of resistant cultivars, with the lowest number of generation breeding cycles precisely through the process of gene pyramiding, in addition to marker-assisted backcrossing, early generation selection and combined MAS. Nevertheless, sequence based markers such as single nucleotide polymorphisms (SNP) and polymerase chain reaction (PCR) based-simple sequence repeat (SSR) or microsatellite markers remain the most preferred marker systems for genetic/genomic studies and crop breeding. Choudhary *et al.* [55] reviewed about 2000 genomic SSR markers but with 30% polymorphic marker frequency as compared to other genera of this tribe [56]. Based on these reports, SSRs exhibit polymorphism in terms of variation in the number of repeat units as revealed by amplification of unique sequences flanking these repeat units, with their co-dominance inheritance suitable for genotyping segregating populations.

Quantitative Trait Loci (QLT)

The identification of DNA markers, genes, and quantitative trait loci (QTLs) associated with particular traits is accomplished through QTL mapping [53]. QTL mapping also allows for a better understanding of the genetic linkages or control and the trait inheritance. This provides insights through the construction of linkage maps that enables breeders to detect specific chromosome that has candidate gene segments for stress resistance or for choosing the best selection strategy composing genetic markers for a specific population. A large number of quantitative trait loci (QTLs) mapping studies for diverse crop species continue to provide an abundance of DNA marker–trait associations [57]. QTL mapping signifies the basis of development of molecular markers for MAS besides the tool's inaccuracies, such as, the replication levels of phenotypic information, population sizes and type, environmental effects, as well as genotyping errors [53, 54].

Using DNA markers for the identification of QTLs was a breakthrough in the characterisation of quantitative traits. Currently, a wide range of QTLs governing various biotic and abiotic stress factors were identified at the genetic level in segregating mapping populations. Chen *et al.* [58] mapped QTLs for heat stress tolerance using second filial generation plants (F_2), recombinant inbred line (RIL) and backcross inbred line (BIL) populations. Additionally, phenotypic variation ranging between 6.27 to 21.29% of the five identified QTLs found on chromosome 5 and 9 were reported by Shanmugavadivel *et al.* [59] to bestow heat stress tolerance in rice, where chromosome 5 was narrowed down even from 23 Mbp to 331 Kbp. Yang *et al.* [60] also reported a broad-sense heritability (h^2_b) for biological nitrogen fixation (BNF) ranging between 0.48 to 0.87% based on the genotype. Moreover, two new QTLs for BNF traits, *qBNF-16* and *qBNF-17* were also identified, following the evaluation of soybean parent lines (with contrasting BNF traits) and 168 $F_{9:11}$ RILs under varied field conditions.

LIST OF ABBREVIATIONS

A/T	Adenine- thymine
BIL	Backcross inbred line
BNF	Biological nitrogen fixation
CO₂	Carbon dioxide
CRISPR-Cas9	Clustered regularly interspaced short palindromic repeats-associated protein 9
DNA	Deoxyribonucleic acid
EMS	Ethylmethane sulfonate
G/C	Guanine- cytosine
MAS	Marker-assisted selection
NaCl	Sodium chlorite
NHEJ	Non-homologous end joining
PCR	Polymerase chain reaction
QTL	Quantitative trait loci
RIL	Recombinant inbred line
ROS	Reactive oxygen species
SNP	Single nucleotide polymorphism
SSR	Simple sequence repeat
UV	Ultraviolet

CONSENT FOR PUBLICATION

Not applicable.

CONFLICT OF INTEREST

The author declares no conflict of interest, financial or otherwise.

ACKNOWLEDGEMENTS

Declared none.

REFERENCES

[1]   Singh U, Singh B. Tropical grain legumes as important human foods. Econ Bot 1992; 46: 310-21.
       [http://dx.doi.org/10.1007/BF02866630]

[2]   Jendoubi W, Bouhabida M, Boukteb A, Beji M, Kharrat M. *Fusarium* wilt affecting chickpea crop.
       Agric 2017; 7(3): 1-6.

[3]   Lavin M, Pennington RT, Klitgaard BB, Sprent JI, de Lima HC, Gasson PE. The dalbergioid legumes
       (Fabaceae): delimitation of a pantropical monophyletic clade. Am J Bot 2001; 88(3): 503-33.

[http://dx.doi.org/10.2307/2657116] [PMID: 11250829]

[4]   Lopez J, Devesa JA, Ortega-Olivencia A, Ruiz T. Production and morphology of fruit and seeds in *Genisteae* (*Fabaceae*) of south-west Spain. Bot J Linn Soc 2000; 123: 97-120.
         [http://dx.doi.org/10.1111/j.1095-8339.2000.tb01208.x]

[5]   Schrire BD, Lavin M, Barker NP, Forest F. Phylogeny of the tribe Indigofereae (*Leguminosae-Papilionoideae*): Geographically structured more in succulent-rich and temperate settings than in grass-rich environments. Am J Bot 2009; 96(4): 816-52.
         [http://dx.doi.org/10.3732/ajb.0800185] [PMID: 21628237]

[6]   Moteetee AN, Van Wyk BE. Taxonomic notes on the genus *Otoptera* (*Phaseoleae, Fabaceae*) in southern Africa. S Afr J Bot 2011; 77: 492-6.
         [http://dx.doi.org/10.1016/j.sajb.2010.10.007]

[7]   Hartmann HT, Kester DE, Davies FT, Geneve RL. Hartmann and Kester's plant propagation: Principles and practices. United States of America: Pearson Edu Ltd. 2014; p. 5.

[8]   Raven JA. The size of cells and organisms in relation to the evolution of embryophytes. Plant Biol 1999; 1: 2-12.
         [http://dx.doi.org/10.1111/j.1438-8677.1999.tb00702.x]

[9]   Taiz L, Zeiger E, Moller IM, Murphy M. Plant physiology and development. London: Sinauer Associates 2015; pp. 756-60.

[10]   Lal MA, Rathpalia R, Sisodia R, Shakya R. Biotic stress. In: Bhatla SC, Lal MA, Eds. Plant physiology, development and metabolism. Singapore: Springer 2018; pp. 1029-95.
         [http://dx.doi.org/10.1007/978-981-13-2023-1_32]

[11]   Dukhovskis P, Juknys R, Brazaityte A, Zukauskaite I. Plant response to integrated impact of natural and anthropogenic stress factors. Russ J Plant Physiol 2003; 50: 147-54.
         [http://dx.doi.org/10.1023/A:1022933210303]

[12]   Fuchs LK, Jenkins G, Phillips DW. Anthropogenic impacts on meiosis in plants. Front Plant Sci 2018; 9: 1429.
         [http://dx.doi.org/10.3389/fpls.2018.01429] [PMID: 30323826]

[13]   Solh MB, Halila HM, Hernandez-Bravo G, Malik BA, Mihov MI, Sadri B. Biotic and abiotic stress constraining the productivity of cool season food legumes in different farming system: Specific example. In: Muchlbaver FJ, Kaiser WJ, Eds. Expanding the production and use of cool season food legumes. Dordrecht: Springer 1994; pp. 219-30.
         [http://dx.doi.org/10.1007/978-94-011-0798-3_12]

[14]   Freitas RMO, Dombroski JLD, De Freitas FCL, Nogueira NW, Pinto JRS. Physiological responses of cowpea under water stress and rewatering in no-tillage and conventional tillage systems. Rev Caatinga 2017; 30(3): 559-67.
         [http://dx.doi.org/10.1590/1983-21252017v30n303rc]

[15]   Mauro RP, Sortino D, Dipasquale M, Mauromicale G. Phenological and growth response of legume cover crop to shading. J Agric Sci 2013; 1-15.

[16]   Haile GG, Tang Q, Sun S, Huang Z, Zhang X, Liu X. Droughts in east Africa: Causes, impacts and resilience. Earth Sci Rev 2019; 193: 146-61.
         [http://dx.doi.org/10.1016/j.earscirev.2019.04.015]

[17]   Zwane EM. Impact of climate change on primary agriculture, water sources and food security in Western Cape, South Africa. Jamba 2019; 11(1): 562.
         [http://dx.doi.org/10.4102/jamba.v11i1.562] [PMID: 31049161]

[18]   Wijewardana C, Reddy KR, Krutz LJ, Gao W, Bellaloui N. Drought stress has transgenerational effects on soybean seed germination and seedling vigor. PLoS One 2019; 14(9)e0214977
         [http://dx.doi.org/10.1371/journal.pone.0214977] [PMID: 31498795]

[19]   Nedumaran S, Abinaya P, Jyosthnaa P, Shraavya B, Rao P, Bantilan C. Grain legumes production, consumption and trade trends in developing countries. ICRISAT Research Program Markets, Institutions and Policies, Working Paper Series 60 2015; 64.

[20]   Amira MS, Qados A. Effect of salt stress on plant growth and metabolism of bean plant *Vicia faba* (L.). J Saudi Soc Agric Sci 2011; 10: 1-15.

[21]   Manchandra G, Garg N. Salinity and its effects on the functional biology of legumes. Acta Physiol Plant 2008; 30(5): 595-618.
[http://dx.doi.org/10.1007/s11738-008-0173-3]

[22]   Sehrawat N, Bhat KV, Sairam RK, Jaiwal PK. Screening of mung bean (*Vigna radiata* L. Wilczek) genotypes for salt tolerance. In J Plant Anim. Environ Sci (Ruse) 2013; 4: 36-43.

[23]   Wang D, Heckathorn SA, Mainali K, Tripathee T. Timing effects of heat-stress on plant ecophysiological characteristics and growth. Front Plant Sci 2016; 7(1629): 1-11.

[24]   Bita CE, Gerats T. Plant tolerance to high temperature in a changing environment: scientific fundamentals and production of heat stress-tolerant crops. Front Plant Sci 2013; 4(273): 273.
[http://dx.doi.org/10.3389/fpls.2013.00273] [PMID: 23914193]

[25]   Guilioni L, Wery J, Tardieu F. Heat stress-induced abortion of buds and flowers in pea: Is sensitivity linked to organ age or to relations between reproductive organs? Ann Bot 1997; 80: 159-68.
[http://dx.doi.org/10.1006/anbo.1997.0425]

[26]   Konsens I, Ofir M, Kigel J. The effect of temperature on the production and abscission of flowers and pods in Snap bean (*Phaseolus vulgaris* L.). Ann Bot 1991; 67: 391-9.
[http://dx.doi.org/10.1093/oxfordjournals.aob.a088173]

[27]   Roy KW. Falcate-spored species of colletotrichum on soybean. Mycol 1996; 88(6): 1003-9.
[http://dx.doi.org/10.1080/00275514.1996.12026742]

[28]   Timko MP, Singh BB. Cowpea, a multifunctional legume. In: Moore PH, Ming R, Eds. Genomics of tropical crop plants Plant genetics and genomics: Crops and models. New York: Springer 2008; Vol. 1: pp. 227-58.
[http://dx.doi.org/10.1007/978-0-387-71219-2_10]

[29]   Moran VC. The phytophagous insects and mites of cultivated plants in South Africa: Patterns and pest status. J Appl Ecol 1983; 20: 439-50.
[http://dx.doi.org/10.2307/2403518]

[30]   Dowley LJ, Grant J, Griffin D. Yield losses caused by late blight *Phytophthora infestans* (Mont. J de Bary) in potato crops in Ireland. Ir J Agric Food Res 2008; 47: 69-78.

[31]   Irwin JAG, Crowford AR, Drenth A. The origins of Phytophthora species attacking legumes in Australia. Adv Bot Res 1997; 24: 431-56.
[http://dx.doi.org/10.1016/S0065-2296(08)60081-6]

[32]   Gaulin E, Jacquet C, Bottin A, Dumas B. Root rot disease of legumes caused by Aphanomyces euteiches. Mol Plant Pathol 2007; 8(5): 539-48.
[http://dx.doi.org/10.1111/j.1364-3703.2007.00413.x] [PMID: 20507520]

[33]   Fritsche-Neto R, Ferreira LR, Ferreira FA, da Silva AA, Do Vale JC. Breeding for weed management. In: Fritsche-Neto R, Borem A, Eds. Plant breeding for biotic stress resistance. Berlin, Heidelberg: Springer-Verlag 2012; pp. 137-64.
[http://dx.doi.org/10.1007/978-3-642-33087-2_8]

[34]   Haile MG, Brockhaus J, Kalkuhl M. Short-term acreage forecasting and supply elasticities for stable food commodities in major producer countries. Agric Econ 2016; 4(17): 1-23.

[35]   Chandra A, Pental D. Regeneration and genetic transformation of grain legumes: An overview. Curr Sci 2003; 84(3): 381-7.

[36]   Christou P. Genetic transformation of crop plants using microprojectile bombardment. Plant J 1992; 2: 275-81.
[http://dx.doi.org/10.1111/j.1365-313X.1992.00275.x]

[37]   Mookkan M. Particle bombardment - mediated gene transfer and GFP transient expression in Seteria viridis. Plant Signal Behav 2018; 13(4): e1441657.
[http://dx.doi.org/10.1080/15592324.2018.1441657] [PMID: 29621423]

[38]   Kato-Inui T, Takahashi G, Hsu S, Miyaoka Y. Clustered regularly interspaced short palindromic repeats (CRISPR)/CRISPR-associated protein 9 with improved proof-reading enhances homology-directed repair. Nucleic Acids Res 2018; 46(9): 4677-88.
[http://dx.doi.org/10.1093/nar/gky264] [PMID: 29672770]

[39]   Ivo NL, Nascimento CP, Vieira LS, Campos FA, Aragão FJ. Biolistic-mediated genetic transformation of cowpea (*Vigna unguiculata*) and stable Mendelian inheritance of transgenes. Plant Cell Rep 2008; 27(9): 1475-83.
[http://dx.doi.org/10.1007/s00299-008-0573-2] [PMID: 18587583]

[40]   Indurker S, Misra HS, Eapen S. Genetic transformation of chickpea (*Cicer arietinum* L.) with insecticidal crystal protein gene using particle gun bombardment. Plant Cell Rep 2007; 26(6): 755-63.
[http://dx.doi.org/10.1007/s00299-006-0283-6] [PMID: 17205334]

[41]   Krishna G, Reddy PS, Ramteke PW, Bhattacharya PS. Progress of tissue culture and genetic transformation research in pigeon pea [*Cajanus cajan* (L.) Millsp.]. Plant Cell Rep 2010; 29(10): 1079-95. [*Cajanus cajan* (L.) Millsp.].
[http://dx.doi.org/10.1007/s00299-010-0899-4] [PMID: 20652570]

[42]   Hinchee MAW, Connor-Ward DV, Newell CA, *et al.* Production of transgenic soybean plants using *Agrobacterium*-mediated DNA transfer. Nat Biotechnol 1988; 6: 915-22.
[http://dx.doi.org/10.1038/nbt0888-915]

[43]   Ko TS, Nelson RL, Korban S. Screening multiple soybean cultivars (MG00 to MG VIII) for somatic embryogenesis following *Agrobacterium*-mediated transformation of immature cotyledons. Crop Sci 2004; 44: 1825-31.
[http://dx.doi.org/10.2135/cropsci2004.1825]

[44]   Raveendar S, Ignacimuthu S. Improved *Agrobacterium*-mediated transformation in cowpea *Vigna unguiculata* L. Walp. Asian J Plant Sci 2010; 9: 256-63.
[http://dx.doi.org/10.3923/ajps.2010.256.263]

[45]   Mehrotra M, Sanyal I, Amla DV. High-efficiency Agrobacterium-mediated transformation of chickpea (*Cicer arietinum* L.) and regeneration of insect-resistant transgenic plants. Plant Cell Rep 2011; 30(9): 1603-16.
[http://dx.doi.org/10.1007/s00299-011-1071-5] [PMID: 21516347]

[46]   Patel M, Dewey RE, Qu R. Enhancing *Agrobacterium tumefaciens*-mediated transformation efficiency of perennial ryegrass and rice using heat and high maltose treatments during bacterial infection. Plant Cell Tissue Organ Cult 2013; 114: 19-29.
[http://dx.doi.org/10.1007/s11240-013-0301-7]

[47]   Ané JM, Zhu H, Frugoli J. Recent advances in Medicago truncatula genomics. Int J Plant Genomics 2008; 2008: 256597.
[http://dx.doi.org/10.1155/2008/256597] [PMID: 18288239]

[48]   Tadege M, Wang TL, Wen J, Ratet P, Mysore KS. Mutagenesis and beyond! Tools for understanding legume biology. Plant Physiol 2009; 151(3): 978-84.
[http://dx.doi.org/10.1104/pp.109.144097] [PMID: 19741047]

[49]   Micke A. Mutation breeding of grain legumes. Plant Soil 1993; 152: 81-5.
[http://dx.doi.org/10.1007/BF00016335]

[50]   McCallum CM, Comai L, Greene EA, Henikoff S. Targeting induced local lesions IN genomes

(TILLING) for plant functional genomics. Plant Physiol 2000; 123(2): 439-42.
[http://dx.doi.org/10.1104/pp.123.2.439] [PMID: 10859174]

[51]   Mostafa GG, Alhamd MFA. Detection and evaluation the tetraploid plants of *Celosia argentea* induced by colchicines. Int J Plant Breed Genet 2016; 10(2): 110-5.
[http://dx.doi.org/10.3923/ijpbg.2016.110.115]

[52]   Mangena P. *In vivo* and *in vitro* application of colchicine on germination and shoot proliferation in soybean. Asian J Crop Sci 2020; 12(1): 34-42. [*Glycine max* (L.) Merr.].
[http://dx.doi.org/10.3923/ajcs.2020.34.42]

[53]   Das G, Patra JK, Baek KH. Insight into MAS: A molecular tool for development of stress resistant and quality of rice through gene stacking. Front Plant Sci 2017; 8(985): 1-9.

[54]   Riday H. Marker assisted selection in legumes. Lotus News Lett 2007; 37(3): 102.

[55]   Choudhary AK, Sultana R, Vales MI, Saxena KB, Kumar RR, Ratnakumar P. Integrated physiological and molecular approaches to improvement of abiotic stress tolerance in two pulse crops of the semi-arid tropics. Crop J 2018; 6(2): 99-114.
[http://dx.doi.org/10.1016/j.cj.2017.11.002]

[56]   Choudhary S, Gaur R, Gupta S, Bhatia S. EST-derived genic molecular markers: development and utilization for generating an advanced transcript map of chickpea. Theor Appl Genet 2012; 124(8): 1449-62.
[http://dx.doi.org/10.1007/s00122-012-1800-3] [PMID: 22301907]

[57]   Collard BCY, Mackill DJ. Marker-assisted selection: an approach for precision plant breeding in the twenty-first century. Philos Trans R Soc Lond B Biol Sci 2008; 363(1491): 557-72.
[http://dx.doi.org/10.1098/rstb.2007.2170] [PMID: 17715053]

[58]   Chen Q, Yu S, Li C, Mou T. Identification of QTLs for heat tolerance at flowering stage in rice. Zhongguo Nong Ye Ke Xue 2008; 41: 315-21.

[59]   Ps S, Sv AM, Prakash C, *et al.* High resolution mapping of QTLs for heat tolerance in rice using a 5K SNP array. Rice (N Y) 2017; 10(1): 28.
[http://dx.doi.org/10.1186/s12284-017-0167-0] [PMID: 28584974]

[60]   Yang Q, Yang Y, Xu R, Lv H, Liao H. Genetic analysis and mapping of QTLs for soybean biological nitrogen fixation traits under varied field conditions. Front Plant Sci 2019; 10(75): 75.
[http://dx.doi.org/10.3389/fpls.2019.00075] [PMID: 30774643]

CHAPTER 2

Genetic Diversity, Conservation and Cultivation of Grain Legumes

Phetole Mangena[1,*], **Samuel Tebogo Posie Peta**[2] and **Arinao Mukatuni**[1]

[1] *Department of Biodiversity, School of Molecular and Life Sciences, Faculty of Science and Agriculture, University of Limpopo, Private Bag X1106, Sovenga 0727, South Africa*

[2] *DSI-NRF Centre of Excellence for Invasion Biology, Department of Botany and Zoology, Faculty of Science, Stellenbosch University, Private Bag X1, Matieland 7602, South Africa*

Abstract: Understanding genetic diversity is essential for achieving genetic improvement and conservation of grain legumes. These crops serve as important pulses grown and consumed all over the world, especially in Africa, south east Asia and America. Legumes serve as a good source of carbohydrates, oil, fibre and proteins. They contain all essential amino acids, nutritionally important unsaturated fatty acids such as linoleic and oleic acids, and mineral elements such as K, Ca, Mg, P and Zn. The seeds contain all important vitamins such as riboflavin, niacin, thiamine, vitamin A precursor ß-carotene and folate. As with other crops like cereals, legumes face reduced genetic diversity, which impact negatively on the production of newly improved varieties showing stress tolerance. Conservation of wild genetic resources and cultivated germplasm will provide genetic materials for future breeding programmes. Genome sequencing libraries and bioinformatics tools could be used to screen and select genotypes with desirable traits, even according to the geographical patterns. The results will be of major importance for conservation genetics and breeding of newly improved cultivars that exhibit high resilience to adverse global climate patterns and plant pathogens.

Keywords: Conservation, Genetic diversity, Grain legumes, Pulse crops.

INTRODUCTION

Legume plants, family *Fabaceae* constitutes about 800 genera and over 15.000 species, making them the third largest family of flowering plants, after *Orchidaceae* and *Asteraceae*. Generally, these are the three largest flowering plant families containing the greatest number of species with 24.000 species for the sunflower family, 20.000 in orchids and 18.000 species estimated for the pea family.

* **Corresponding author Phetole Mangena**: Department of Biodiversity, School of Molecular and Life Sciences, Faculty of Science and Agriculture, University of Limpopo, Private Bag X1106, Sovenga 0727, South Africa; Tel: +2715-268-4715; E-mails: mangena.phetole@gmail.com & phetole.mangena@ul.ac.za

The pea family was reported by Smykal *et al.* [1] as an extremely diverse taxon with worldwide distribution of wild and domestically cultivated species. These plants have a broad range of life forms, composing of arctic alpine plants, temperate or tropical annual shrubs and herbs, as well as equatorial giant trees. Some of these species have an undesirable growth, while others are used as major grain crops in agriculture. Grain legumes used for agricultural purposes include the common bean, pea, chickpea, cowpea, faba bean, pigeon pea, soybean and grass pea, *etc.*

These plants occupy a considerably high rank in agriculture, and all species contain a large number of varieties in which many of them are bred and developed for both subsistent and commercial farming [2]. Legume crops serve as essential sources of high quality proteins (estimated to be above 40%), minerals (like phosphorus, copper, magnesium and manganese), lipids (ranging about 20%), water soluble carbohydrates (over 20%), fibre, and various vitamins (*e.g.* thiamine, folic acid, riboflavin, vitamin B6 *etc.*) that are mostly contained within the seeds. All these seed constituents are essential for energy generation, primary and secondary metabolite biosynthesis and serve as cofactors for carboxylation/oxidation of residues involved in blood coagulation, bone metabolism, prevention of vessel mineralisation and regulation of the various cellular functions in humans [3].

The United Nations, in collaboration with several pulse/grain organisations and research institutes, promoted and recommended legume cropping and legume-protein-based food consumption as the most affordable and readily available alternative for meat protein [4]. With many regions experiencing global warming due to climate change and the increased threat of severe drought, which impacts negatively on agricultural production, legume crops could be easily genetically manipulated to counteract these environmental stress effects. The African continent remains home to most of the world's poorest population, which heavily relies on inefficient agricultural systems for survival. Thus, many of these crops (Table **1**) are considered to be ideal for cultivation in semiarid and arid regions as they exhibit an extensive tap root system.

This kind of root system and architecture is critical for helping plant roots reach soil moisture content held in deeper soil layers and making them drought resistant, when compared to fibrous roots [5, 6]. However, conservation of the already existing genotypes, and the development of new genetic characteristics are required for improved growth and productivity under the changing adverse environmental conditions. The existing germplasm needs to be preserved because many of the modern agricultural practices tend to reduce the genetic diversity of the plants utilised. This is vitally important even for the development of new genetic resources through plant breeding to develop varieties that are disease resistant in the event of crop losses as a result of environmental changes, increasing pathogen virulence and unavailability of arable land for agriculture. This chapter focuses on the need to improve genetic diversity, conservation and cultivation of legume crops for expanding production, particularly when confronted with biotic and abiotic stress factors.

Table 1. Major domesticated legume plant species and their proximate essential constituents abundant in the seeds. Information includes indications of both widely cultivated regional and non-regional species, and their widely used common names.

	Grain Species	Common Name	Oil	Protein (% Dry Weight)	Sugars
Regional Asiatic Pulses					
1.	Vigna aconitifolia	Moth bean	2.1-2.7	30-46	10-20
2.	*Vigna angularis*	Azuki bean	1.3-1.5	20-33	15-33.6
3.	*Vigna mungo*	Urd bean	2.1-2.7	20-24.5	35-40
4.	*Vigna umbellate*	Rice bean	2.4-3.3	19-24	30-35
None-Regional Based Varieties					
5	*Cajanus cajan* L.	Pigeonpea	0.6-3.8	18-26	51-58
6	*Cicer arietinum* L.	Chickpea	1.0-6.0	20-30	30-40
7	*Glycine max* L.	Soybean	20-22	20-40	5-10
8	*Lens culinaris* L.	Lens	1.0-2.8	20-36	40-55
9	*Phaseolus vulgaris*	Common bean	1.5-2.0	16-33	33-40
10	*Pisum sativum* L.	Peas	0.0-1.0	8.7-11.1	4.0-20
11	*Vigna radiata*	Mung bean	1.2-1.6	14-21	20-32
12	*Vicia faba* L.	Faba bean	1.0-3.4	20-28	42-47
13	*Vigna unguiculata* L.	Cowpea	2.1-2.3	22-26	56-68

DISTRIBUTION AND DIVERSITY OF LEGUMES

Grain legumes are valuable crops, providing nutritious foods for the expanding world populations and becoming increasingly important under the changing climate, which affects both crop and animal production. The value of legumes in terms of nutrition and body health was indicated by Marga and Haji [7] to be of highest significance in the diet of many people in the developing countries, thus being referred to as the poor man's meat. These grain crops also serve as a major component of animal feed and have an important role to play in the environment. However, the global diversity and distribution of agricultural legume species contain a narrow genetic pool, remain less documented and underexplored for cultivation to meet human/animal needs. The information regarding their world-wide distribution, as indicated in Fig. (**1**), is required to identify potential species with highly enhanced vegetative growths, stable cell architecture and yields [8, 9]. Grain legumes are required to complement cereals as affordable sources rich in proteins, fibre and carbohydrates, especially for animal and human diet in developing countries [8].

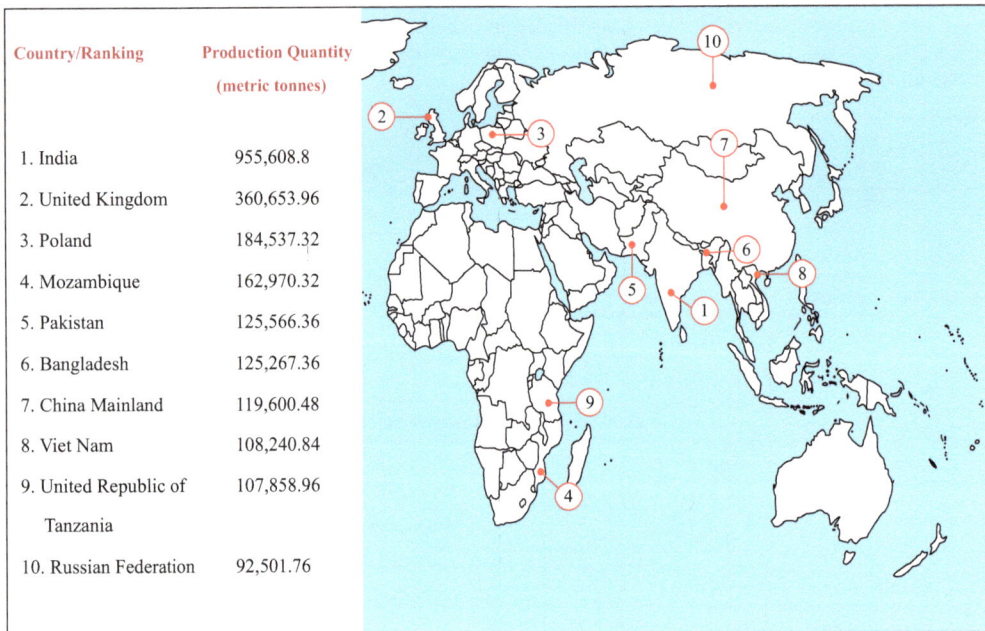

Country/Ranking	Production Quantity (metric tonnes)
1. India	955,608.8
2. United Kingdom	360,653.96
3. Poland	184,537.32
4. Mozambique	162,970.32
5. Pakistan	125,566.36
6. Bangladesh	125,267.36
7. China Mainland	119,600.48
8. Viet Nam	108,240.84
9. United Republic of Tanzania	107,858.96
10. Russian Federation	92,501.76

Fig. (1). Illustration of worldwide distribution and production of legume species used as grain crops. The presented data belongs to only an overall fraction of legume species used as major commodities for the top 10 countries serving as the highest legume crop producers, according to FAOSTAT [9].

Fabaceae remains one of the flowering plant taxa with the largest number of plant species that are widely distributed around the world. The species grow across a wide range of climatic conditions and soil types (Fig **1**). Such regions include the Equatorial Rainforest region, where the climate is fully humid, Savannah climate with very dry summers and winters. As well as the warm temperate climate in countries like the United Kingdom, Poland, and China Mainland [9]. These regions exploit only a fraction of these species falling in this Fabaceae group for agricultural purposes. This remains the case, despite a large morphological and growth response diversity that exists in this family, with more than 18,000 documented species. For decades, *Pisum sativum* (peas), *Glycine max* L. (soybeans), *Phaseolus* spp. (common bean), and *Vigna* spp. (cowpea) (Fig. **3**) covered a range of about 70–78% of the area under legume cultivation in Europe,

Oceania, and Africa. In North and South America, more than 76% of the area allocated for pulse legumes remain utilised largely for the cultivation of soybeans [6, 8, 9].

In terms of production volumes, the legumes are second to cereals (Poaceae) that are also most important for providing large quantities of stored carbohydrates in the form of starch. However, due to the ability to fix nitrogen, legumes exhibit a high tolerance to various environmental conditions than cereals [10]. The large distribution of legume species world-over will therefore, assist researchers and crop breeders to identify cultivars with high stress tolerance and yield production. For the past two decades, many laboratories and field experiments have assessed ways to improve agronomic and environmental performance of grain legumes. These assessments are conducted worldwide, and findings vary due to field specificity, growing seasons and genotype as function of the grain's adaptability/environmental fitness, soil type, temperature, and climate. Examples of the selected and widely cultivated legume species are discussed below.

Phaseolus vulgaris

Common bean (*Phaseolus vulgaris* L.) is considered one of the most widely distributed crops among the annual grain legumes (Fig. **3A**). This species has the broadest range of genetic resources [11], achieving the highest diversity of landraces that are considered economically and ecologically valuable to sustain their own distribution and genetic diversity. Many of the landraces are cultivated privately for subsistence and by small scale farmers for commercial purposes. Common bean has both early and late maturing varieties with either determinate or indeterminate growth forms. This species is well-adapted to local weather patterns, shows tolerance to abiotic stress and is highly susceptible to biotic stress factors [12]. However, this bean still needs to be artificially bred to introduce new varieties with increased resistance to insect pests and pathogenic agents such as bacteria, fungi and viruses. Many small-scale farmers have so far, complemented natural section by artificially producing heterogenous landrace varieties. This allows them to select high yielding varieties and achieve specific traits, ultimately leading to the planting of different populations of the same landraces. These populations serve as crucial sources of germplasm that may also be preserved to maintain the diversity and will be useful for genetic improvements of the crops [13]. Stoilova *et al.* [11] reported the use of cluster analysis to evaluate landraces suitable to be used as genetic resources to identify important phenotypic parameters to be considered in future common bean breeding activities. The procedure assessed several agronomical, morphological and phenological characteristics in Portuguese and Bulgarian landraces of *Phaseolus vulgaris* L.

The study revealed four accessions [number 99E059(PG), 99E0123(BG), PH23(PT) and PH2(PT)] that demonstrated a short life cycle, erect habit and high number of pods and seeds per plant as the desired special interest traits for breeding purposes. Furthermore, the landraces were found to contain similar unique characteristics (Fig. **2**) and largely distributed among regions in Portugal and Bulgaria. Similar studies were conducted by Asfaw *et al.* [14] and Zhang *et al.* [15] in order to increase the genetic diversity of landraces available for breeding programmes. However, common bean cultivation by farmers may continue to influence the crop's genetic diversity as a result of the artificial selections and agroecological conditions involved during farming practices. These dynamic effects are exacerbated by the fact that farmers produce plant seed materials themselves or source them from fellow farmers and from the local seed suppliers, than continuously purchasing their seeds from breeding companies which guaranteed seed quality and purity.

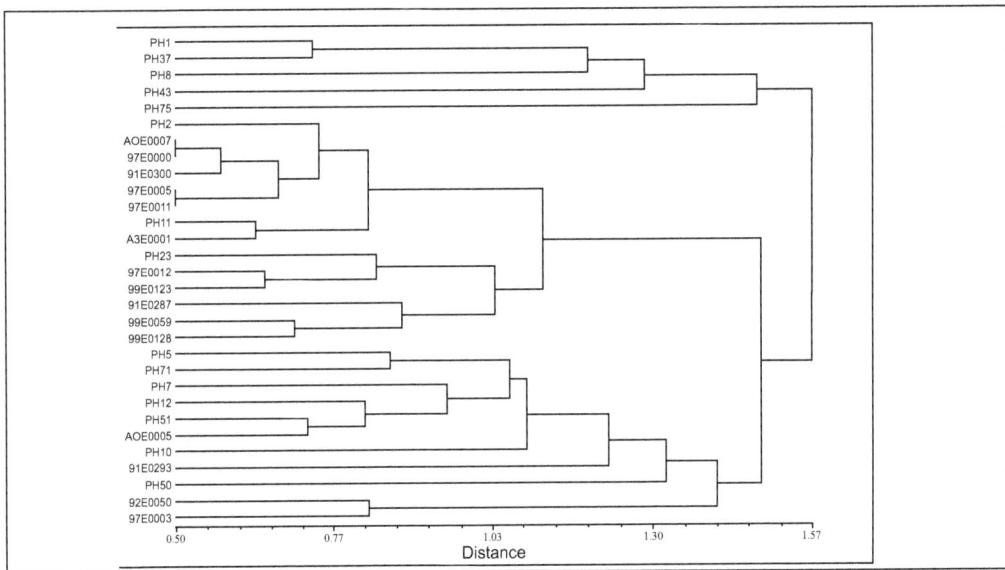

Fig. (2). Dendrogram illustration representing taxonomical relationship among the landraces found in Portugal and Bulgaria by Stoilova *et al.* [11].

Cicer arietinum

Chickpea (Fig. **3B**) is one of the most important cash crops grown in the highlands of many countries worldwide. This crop currently ranks third among pulses, with the mean annual production volume of about 11.5 million tons. Chickpea is currently grown at about 14.6 million hectares worldwide, with most of the production focussed on India, followed by Australia, Canada, and Argentina. In contrast to other popular legume species like soybean, the developing countries share the most fraction of world chickpea production. An estimated total of 95% of the production area, processing and consumption of chickpea is forecasted in those countries. Furthermore, production per unit area has steadily increased in the past five decades with almost 7 kg/hectare recorded per annum. The cultivation resulted with an estimated 2.3 million tons of chickpea entering the world market annually to supplement the production values of countries that are unable to meet the demand through domestic production [7].

However, chickpea's diversity is very narrow, constituting only two main cultivated varieties. Kabuli types (with usually large, smoother and generally light coloured seeds) and Desi types (with various colours, small angular and sometimes spotted seeds) are the available commercial varieties in the gene pool. Like *Phaseolus vulgaris* and other grains, the production of chickpea remains low, probably due to lime induced iron deficiency along with biotic and abiotic stress factors [17]. Other authors attribute these low rates to the lack of improved smart agricultural technology and the lack of highly viable seeds. Competition with other major legume crops, as well as wheat and cotton, were also highlighted. Various biochemical, morphological and molecular markers were also evaluated using cluster analysis to characterise and compare local as well as exotic varieties for exploration in developing new breeding lines [16].

Most populations prefer to consume chickpea directly as food or in various processed home-based forms of foods or as feed in animal farming. Chickpea, like other legumes is often grown as a rotation crop with cereals because of its role in nitrogen fixation. Additionally, the development of high-yielding cultivars showing resistance to diseases are still required to counteract the slow growth experienced due to stress.

Cajanus cajan

As in chickpea (*Cicer arietinum*), pigeonpea (*Cajanus cajan* L.) is severely affected by the continued loss in genetic resources required to enhance its population diversity. Most annual crop species that are self-pollinated or partially out crossing species [18] remain susceptible and vulnerable to genetic depression. This species, together with other related genera currently experience reduced biological fitness to survive and perpetuate its genetic materials to produce diversified populations. Thus, the continuous failure to successfully cross wild relatives and landraces as a source for increasing diversity in the breeding materials will continue to cause some difficulties in the breeding of *Cajanus cajan* against biotic stress. Pigeonpea (Fig. **3C**) as a member of the tribe *Phaseoleae* that constitutes several genera including *Phaseolus*, *Vigna* and *Lablab* contain genes that have not as yet been transferred and integrated into the breeding systems to produce new hybrid combinations. According to Schierenbeck and Norman [19] there are secondary and tertiary gene pools that may contribute to crop improvement, especially those emanating from *Cajanus* wild relative or other closely related plant species. These gene pools require extensive work to cross them into the cultivated gene pool. The cultivated pigeonpea has been presumed to have evolved from interspecific hybridisation of *Cajanus cajanifolia* and *Cajanus scarabaeoides* [20]. Although, these species are different, they contain the same genus unlike those that may not be from a similar taxon. Pigeonpea is widely grown in the subcontinent of India, southern and eastern parts of Africa, southern Asia and the central as well as west America. Amongst these regions, India still account for more than 70% of pigeonpea production in the global market [21]. About sixteen wild species of pigeonpea were reported to exist in India, while only one close relative (*C. kerstingii*) was found to be endemic to Africa, while *C. scarabaeoides* (L.) was introduced to the African continent recently. This made Africa the second region after India with a great landrace populations distribution and diversity in pigeonpea, especially in East Africa, according to Singok *et al.* [21].

Vigna radiata

Conserving the germplasm diversity and relationship among breeding materials is critical for future crop improvements. Wild relatives of crops are essential reservoirs of natural biodiversity, constituting gene pools with tolerance to abiotic stress, disease resistance and other characters normally absent in the breeding systems [19]. Cultivated mung bean (Fig. **3D**) also suffers from low levels of cultivatable genetic diversity, which is probably the reason behind its relatively low productivity in the global market. Although, to date, it is still one of the most important grain legumes, particularly in Asia, with about 2.5–3.1 million tons per year [22]. However, Sangiri *et al.* [22] indicated that this crop was first domesticated in India where wild mung bean remains widely distributed. The report indicated that this claim was based on archaeobotanical records. Mung bean is largely grown in China, with an estimated 1-million hectare of cultivated land.

According to Tomooka *et al.* [23], it belongs to the Asian genus *Vigna*, subgenus *Ceratotropis* which has been divided into *Ceratotropis*, *Angulares* and *Aconitifoliae*. This grain legume species contains some of the largest conserved genetic resources of cultivated landraces with an estimated 5600 accessions found at the World Vegetation Centre in Asia [23]. Mung bean is also cultivated as both food and fodder crop. Consumers prefer varieties with large to medium seed sizes for food and it is considered as a quality pulse due to its excellent digestibility and rich proteins [24]. Improvement and maintenance of such genetic diversity are crucial for the production of enhanced cultivars that will progressively increase its productivity and permit the development of cost-effective conservation strategies. Currently, studies examine genetic diversity of mung bean population using morphological markers and simple sequence repeats (SSRs).

Vigna unguiculata

Cowpea, *Vigna unguiculata* (L.) is considered an important indigenous grain legume crop in sub-Saharan Africa, preferred for consumption as both green vegetable and dry seeds (pulse). In addition, immature seed pods, green leaves and roots are among the edible and utilised plant parts. This crop is cultivated mainly in the dry savanna areas as an intercrop with maize, sorghum and millet. Cowpeas are rich in protein, carbohydrate, fibre and vitamins (Fig. **3E**). The proteins consist of 90% water-insoluble globulins and 10% water-soluble albumins, with isoleucine, lysine, glycine, tryptophan, alanine, proline, glutamine and valine being among the full complement of amino acids found in mature dry seeds. Mfeka *et al.* [25] reported cowpea as a neglected and underutilised crop species among the major legumes.

However, the crop successfully thrives in most arid regions because of its ability to survive hotter condition, alkaline and low fertile soils. Africa contribute about 68% in the global production of cowpea, followed by Brazil (17%), Asia (3%), and USA (2%). The remaining countries in the world contribute at least a total of 10% combined [25]. Cowpea also contain the most sensitive collection of genetic resources located at the International Institute of Tropical Agriculture (IITA) in Nigeria. The germplasm collection contain about 15,000 cultivar accessions from 1975 till to date. All accessions exhibit tolerance to drought, high pH, nematodes, some viruses and weeds. But there is still limited information on its cultivation, seed handling and commercialisation. Reports indicated that, lack of interest on cowpea by crop scientists and funding agencies consequently led to inadequate information regarding its cultivation and nutritional value [25, 26].

Crop improvement is required to deal with large host of diseases, pests and insects that are responsible for massive losses of yields recorded. This remains the case despite a wide range of benefits associated with this crop. In Africa, cowpea serves as a crucial cheap source of proteins compared to other expensive sources such as meat and fish proteins. It plays a critical role in the diets of rural African communities where it is consumed as meshed cowpea or in combination with starchy foods like pap or maize [26]. The crop is adapted to diverse range of climate and soil types, seeds remain viable for several years but, it still gains industrial importance at a very gradual rate [4, 12, 25, 26].

Vicia faba

Despite the significant gains in the global grain legume production market during the past two decades, faba bean (*Vicia faba* L.) (Fig. **3F**) productivity has recently declined. This is attributed to the lack of high yielding cultivars with tolerance to drought and pathogenic resilience. The global cultivation and yield production of faba bean acreage showed significant decline from 3.7 million hectares to 2.1 million hectares between 1980 and 2014 [27]. Specific countries continue to record variable yields, however, productivity per area for some has slightly increased due to the few newly introduced cultivars with reduced susceptibility to biotic and abiotic stress. This crop has been extensively used as a model crop for ascertaining and comparing the effects of mutagenic agents to induce genetic mutation. Filippetti and Ricciardi [28] reported the development and enlargement of useful genetic variability in *Vicia faba* L. The study provided insights into some induced mutations that contribute to improved yield of special interest. This provide baseline information for breeders to create genetic resources required for development of new faba bean plant types. Faba bean already has a number of inheritable and recognisable varieties that exist, which can be employed as markers in cytogenetic analysis and possibly in genetic transformation [28, 29]. It is considered the fourth most important pulse crop in certain regions of the world after soybean, peas and chickpea. In several Mediterranean countries, where faba bean is consumed as a traditional fresh vegetable food crop, the competition with high yielding and input-responsive crops like soybean in irrigated areas has negatively impacted its area and production.

The lack of scientific interest in its morphological and molecular characteristics has supported this persistent form of crop marginalisation. Similar observations were also demonstrated for the cultivation of *Vigna unguiculata* (L.). Over the past decades, faba bean has been used for human consumption due to its highly nutritious content. This crop has high protein content (up to 35% in dry seed) and serves as an excellent source of potassium, calcium, magnesium, iron and zinc [30]. Additionally, it contains essential bioactive compounds such as carotenoids and polyphenols, as well as higher quantities of carbohydrates [31]. However, the high-yield production and improvement genes are still required to sustain the chemical composition that is strongly influenced by a variety of biotic and environmental plant-stress inducing conditions.

Pisum sativum

Peas (*Pisum sativum* L.) is a frost tolerant, cool weather pulse crop and the leading and most popular legume vegetable world over. This crop is a self-pollinated diploid ($2n = 14$) annual pulse and it is highly valued for its high-quality vegetable proteins. It is consumed through grounding of dry mature seeds (Fig. **3G**) into fine powder and mostly used in soap production. Peas are widely cultivated in cooler temperate zone and tropical highlands regions of the world [32]. According to Kindie *et al.* [32] and FAO [27], the crop is grown in a wide range of soil types from light sandy loams to heavy clay but does not tolerate saline and waterlogged soil conditions. In Ethiopia, peas serve as an important vegetable crop and cover about 25149.69 hectares, with an estimated annual production volume of 2.1 million tons [32, 33].

The crop plays a significant role in soil fertility and improving nitrogen content of the soil. It has also gained a unique position in cereal-based cropping system [33]. Other researchers consider it "environmentally friendly and economically feasible from soil improvement point of view" [33, 34]. Kosev and Pachev [35], Grant *et al.* [36] and Krejci *et al.* [37] exploited the possibilities of introgression of desirable gene segments using various DNA transfer methods for genetic improvement of peas. These studies, including various others, clearly presented that resistance genes of interest from different foreign sources could be incorporated into pea genome to counteract the major yield limiting constraints. Examples of such factors include the problematic pea attacks by aphids, pea weevil, powdery mildew, ascochyta blight and other drought-induced stress factors. Krejci *et al.* [37] transferred some DNA segments using *Agrobacterium tumefaciens*-mediated T-DNA transfer using immature embryogenic segments from germinated pea seeds. This was a significant break from the other conventional breeding methods that would require many breeding cycles before desirable genes of interest can be expressed to produce new varieties of pea plants. However, novel breeding strategies such as Advanced Back-cros--Quantitative Trait Loci (AB-QTL) recommended by Singh *et al.* [38] for interspecific hybridisation of *Cicer* spp. to improve productivity and resistance to disease may play a significant role in the genetic improvement of *Pisum* spp. against biotic and abiotic stress.

Lens culinaris

Lentil (*Lens culinaris*) is an ancient leguminous seed crop believed to be indigenous to south western Asia and the Mediterranean region (Fig. **3H**). Lentil is one of the earliest domesticated plant species, with archaeobotanical records dating it back to 7.500 to 6.500 BC. Reports show that it was given the scientific name *Lens culinaris* in 1787 by a German botanist and physician Medikus [39]. Despite being regarded as the oldest cultivated legume plant [40], the genetic diversity and population structure of lentil germplasm collection is still being assessed by various researchers. This is mainly done to support the crop's conservation and genetic enhancement strategies. Khazaei *et al.* [41] used a cultivated lentil collection consisting of 352 accessions originating from 54 diverse countries to estimate its genetic diversity and structure using single nucleotide polymorphism (SNP) markers. Findings in this study indicated a considerably conserved genetic diversity that can be used for breeding purposes.

Furthermore, the germplasm appeared to possess a very narrow genetic pool for certain regions (South Asia and Canada). Consequently, these genetic gaps must be filled for lentil to achieve its diversified role in the farming systems, nutritional security and legume research like other prominent and widely cultivated pulses like soybean. Currently, the big producers of lentil only produce a combined total of 4.2 million tons (Bangladesh, Canada, China, India, Iran, Nepal, Syria and Turkey) with a total cultivated area of about 4.6 million hectares [27]. However, like other legume crops, several diseases severely affect lentil causing major yield losses. The most common diseases reported so far include *Fusarium* wilt caused by *Fusarium oxysporum*, rust caused by *Uromyces fabae* and ascochyta blight as in faba bean caused by *Ascochyta lentis* [39]. However, lentil continues to play a critical role in the livelihoods of many human populations and animals. The crop is continuing to providing services to human lives and animal feed even though its agricultural practices and breeding research are not yet fully clarified.

Fig. (3). List of some of the major grain legume species cultivated for commercial purposes and industrial processing. **(A)** Common bean (*Phaseolus vulgaris*), **(B)** Chickpea (*Cicer arietinum* L.), **(C)** Pigeon pea (*Cajanus cajan* L.), **(D)** Mung bean (*Vigna radiata* L.), **(E)** Cowpea (*Vigna unguiculata* L.), **(F)** Faba bean (*Vicia faba L.*), **(G)** Pea (*Pisum sativum* L.), **(H)** Lentils (*Lens culinaris L.*), **(I)** Soybeans (*Glycine max* L.).

Glycine max

Soybean (*Glycine max* L.) is the most important legume and oilseed crop that plays an irreplaceable role towards sustainable agriculture and maintenance of soil fertility through biological nitrogen fixation via a highly specialised symbiotic relationship with *Rhizobia* bacteria (Fig. **3I**). A sophisticated signalling exchange mediated by phytohormones and glycopeptides (Fig. **4**) control soybean root infection by the bacteria, and induce the formation of novel organs termed nodules. Among all the legume crop species, soybean carryout this interaction highly effectively, allowing the nodules to be fully colonised by the bacteria and provide them with an ideal habited to convert atmospheric nitrogen into nitrogen-based compounds usable by plants. This best studied mutualistic relationship among plants-microbe interactions is responsible for providing nitrogen utilised during plant growth, and it get replenished once the plant dies and subsequently decomposes.

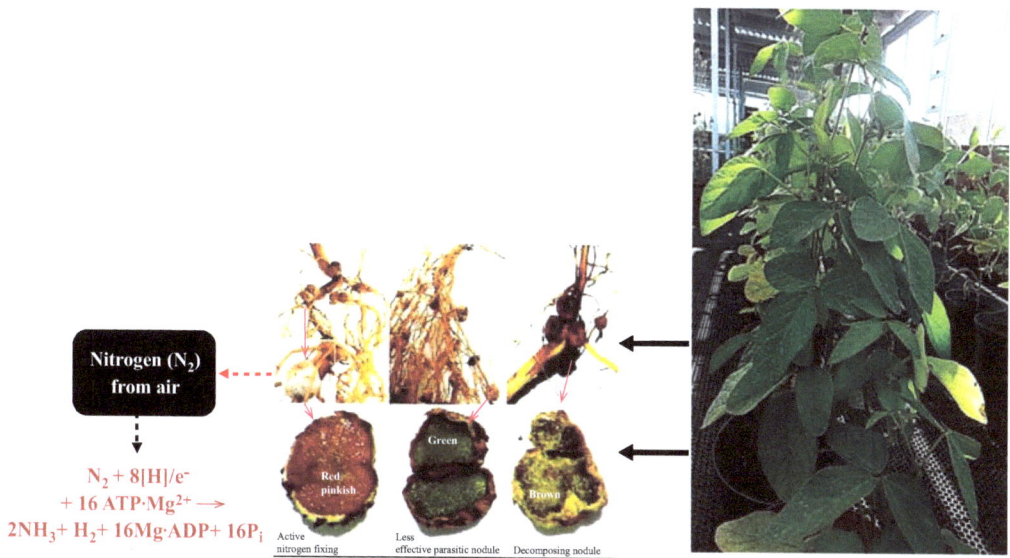

Fig. (4). Nodulation in soybean plant. Nodule formation is regulated by phytohormones and glycopeptides exchange and signalling to ensure optimal plant growth as described by Tatsukami and Ueda [42]. The presented root nodules indicate active nitrogen fixing, less effective and decomposing nodules under different water-deficit conditions [6].

According to Ferguson [43], this process produces approximately 200 million tons of fixed nitrogen used by plants to increase yields and improve soil fertility. Furthermore, agricultural benefits associated with nodulation include immense amount of proteins and oil synthesised in legume crop. Soybean is a major source of protein (40%), oil (20%) and carbohydrates (35%) due to its rapid rise in commercial value at a global scale. Currently, it serves as an important cash crop with total production of over 315.5 million metric tons for 2015–2016 only, with an occupation of around 6% of the world's arable land under its cultivation [44]. The United States, Brazil, Argentina, China, India, Paraguay, and Canada have meanwhile remained the leading producers of soybean with 35, 31, 17, 4, 3, 3 and 2%, respectively. Collectively, legumes have a global production value of over $200 billion per annum, only second to cereal crops, and are cultivated on 12–15% of the world's available arable land.

While soybean is being extensively used as an oilseed and animal feedstuff in many parts of the world, attempt to improve its growth and yield under stress conditions is underway in many laboratories worldwide. Genetic engineering and breeding tools have both been used by researchers to improve the growth of soybean, especially under drought stress. There has been a major global emphasis on genetic improvement of soybean because of its importance, and researchers in North America as well as China, have continued with soybean breeding research for more than 65 years. Its distribution and population structure show a greater array of morphological diversity which reflect the ancient famers' role in developing the current reservoir of available genetic resources since its domestication 3000 years ago [45]. Research still clearly demonstrates that the currently available knowledge about the crop, although valuable, it is still inadequate to accelerate improved utilisation, conservation and management of germplasm collections that are at present time in existence.

LEGUMES CONSERVATION AND PERSPECTIVES

Grain legumes are among the largest flowering plant families worldwide, consisting of over 800 genera and 20,000 species. Plants of this family are uniquely distinguished by the legume fruit, which is simultaneously used as a family name and includes cowpea, lentil, bean and chickpea. This group undoubtedly provide quality essential nutrients such as oil and protein, contributing to food and health security [46]. Due to their several functions, legumes have the potential for conservation agriculture, whereby they function as growing crops, intercropping system or for improving soil fertility [47]. The recent rise in food prices has led to an increase in demand for legumes worldwide. This increasing demand, especially in developing countries, will drive significant changes in the enhancement, conservation and utilisation of the available genetic diversity of legume crops.

The conservation of natural genetic resources may be the only essential tool in maintaining the biological diversity of legumes to continue offering food, fibre, forestry and medicinal services. This is necessary when only a few germplasm populations are available as stocks or the collection of plants and seeds from ancestral wild plants for hybridisation is no longer possible or should be minimised to avoid extinction. Many organisations and agencies have been established that are involved in developing and supporting efforts to conserve native plants through the use of germplasm preservation and *ex situ* and *in situ* conservation. Conservation through *ex situ* means safeguarding biological components outside their place of origin and *in situ* conservation is protection within the natural habitat [48].

The human population is constantly increasing, and food security is under threats, as a result, humans are negatively affected by continuing to live in great poverty. Thus, conservation of genetic materials in legumes as a major affordable source of the much-needed proteins and other valuable bioactive compounds remain important to enable future breeding programmes in order to secure food and agricultural productions [46]. Integrated policy development and implementation of conservation practices are still required to maintain the legume genetic resources for both environmental, livelihood and agroeconomic benefits [47].

CONCLUSION

Grain legumes are a significant source of income and nutrition for billions of poverty stricken populations, particularly in developing countries worldwide [49]. Although remarkable progress has been made in the production of these crops in the past several decades, feed and food legume crop species still possess the potential for plant improvement to confer resistance against several biotic and abiotic stress factors. Many consumers will then realize that the benefits of these crops are exploited to their fullest potential. Soybean undoubtedly remains the chief produced crop among all the legumes, and has considerable competitive agronomic performance, highest protein content, bioactive ingredients and favourable protein quality for livestock feeding. Nevertheless, preservation of this crop including the available natural genetic resources and domestication-associated selection according to local adaptability is a prerequisite for all food, health and ecological services provided by these legumes. The detection of wider molecular diversity and genetic base using gene sequencing would be relevant in establishing distinctness and selection of desirable traits required for breeding systems. The assembly of genetic resources and potential improvements using biotechnology tools will be useful to facilitate domestication and targeted use of germplasms for agronomic trait improvement in most crop legumes.

LIST OF ABBREVIATIONS

AB-QTL　　Advanced back-cross-quantitative trait loci

DNA　　Deoxyribonucleic acid

FAOSTAT　Food and Agriculture Organisation of the United Nations Statistics

IITA　　International Institute of Tropical Agriculture

T-DNA　　Transfer-deoxyribonucleic acid

SNP　　Single nucleotide polymorphism

SSRs　　Simple sequence repeats

Ca　　Calcium

K　　Potassium

Mg　　Magnesium

Zn　　Zinc

CONSENT FOR PUBLICATION

Not applicable.

CONFLICT OF INTEREST

The author declares no conflict of interest, financial or otherwise.

ACKNOWLEDGEMENTS

Declared none.

REFERENCES

[1]　Smykal P, Cogne CJ, Ambrose MJ, *et al.* Legume crops phylogeny and genetic diversity for science and breeding. Crit Rev Plant Sci 2014; 34(1-3): 43-104.
[http://dx.doi.org/10.1080/07352689.2014.897904]

[2]　Mangena P. *In vivo* and *in vitro* application of colchicine on germination and shoot proliferation in soybean. Asian J Crop Sci 2020; 12: 34-42. [*Glycine max* (L.) Merr.].
[http://dx.doi.org/10.3923/ajcs.2020.34.42]

[3]　Foster-Powell K, Holt SHA, Brand-Miller JC. International table of glycemic index and glycemic load values: 2002. Am J Clin Nutr 2002; 76(1): 5-56.
[http://dx.doi.org/10.1093/ajcn/76.1.5] [PMID: 12081815]

[4]　Vollmann J. Soybean versus other food grain legumes: A critical appraisal of the United Nations international year of pulses 2016. Die Bodenkultur: J Land Man Food. Environ 2016; 67(1): 17-24.

[5]　Kant C, Pandey C, Verma S, Tiwari M, Kumar S, Bhatia S. Transcriptome analysis in chickpea (*Cicer arietinum* L.) application in study of gene expression, non-coding RNA prediction, and molecular marker development. In: Marchi F, Cirrillo P, Mateo EC, Eds. Application of RNA-seq and omics strategies: From microorganisms to human health. London, United Kingdom: IntechOpen 2017; pp. 245-63.

[http://dx.doi.org/10.5772/intechopen.69884]

[6]   Mangena P. Water stress: Morphological and anatomical changes in soybean (*Glycine max* L.) plants. In: Andjelkovic V, Ed. Plant, abiotic stress and responses to climate change. London, United Kingdom: IntechOpen 2018; pp. 9-31.
[http://dx.doi.org/10.5772/intechopen.72899]

[7]   Merga B, Haji J. Economic importance of chickpea: Production value and world trade. Cogent Food Agric 2019; 5: 1615718.
[http://dx.doi.org/10.1080/23311932.2019.1615718]

[8]   Cernay C, Pelzer E, Makowski D. A global experimental dataset for assessing grain legume production. Sci Data 2016; 3: 160084.
[http://dx.doi.org/10.1038/sdata.2016.84] [PMID: 27676125]

[9]   FAOSTAT data. Food and Agriculture Organisation of the United Nations. Date accessed December 2019. faostat3.fao.org//home/E/

[10]   Mangena P. A simplified *in-planta* genetic transformation in soybean. Res J Biotechnol 2019; 14(9): 117-25.

[11]   Stoilova T, Pereira G, Tavares de Sousa MM, Carnide V. Diversity in common bean landraces (*Phaseolus vulgaris* L.) from Bulgaria and Portugal. J Cent Eur Agric 2005; 6(4): 443-8.

[12]   Harlan JR. Crops and man. 2nd ed. Madison, Wisconsin, USA: Amer Soc Agron Crop Sci Soc Amer 1992; p. 284.
[http://dx.doi.org/10.2135/1992.cropsandman]

[13]   Bellucci E, Bitocchi E, Rau D, *et al.* Population structure of barley landrace populations and gene-flow with modern varieties. PLoS One 2013; 8(12): e83891.
[http://dx.doi.org/10.1371/journal.pone.0083891] [PMID: 24386303]

[14]   Asfaw A, Blair MW, Almekinders C. Genetic diversity and population structure of common bean (*Phaseolus vulgaris* L.) landraces from the East African highlands. Theor Appl Genet 2009; 120(1): 1-12.
[http://dx.doi.org/10.1007/s00122-009-1154-7] [PMID: 19756469]

[15]   Zhang X, Blair MW, Wang S. Genetic diversity of Chinese common bean (*Phaseolus vulgaris* L.) landraces assessed with simple sequence repeat markers. Theor Appl Genet 2008; 117(4): 629-40.
[http://dx.doi.org/10.1007/s00122-008-0807-2] [PMID: 18548226]

[16]   Ahmad Z, Muntaz AS, Nisar M, Khan N. Diversity analysis of chickpea (*Cicer arietinum* L.) germplasm and its applications for conservation and crop breeding. Agric Sci 2012; 3(5): 723-31.
[http://dx.doi.org/10.4236/as.2012.35087]

[17]   Upadhyaya HD, Bramel PJ, Singh S. Development of a chickpea core subset using geographic distribution and quantitative traits. Crop Sci 2001; 41: 206-21.
[http://dx.doi.org/10.2135/cropsci2001.411206x]

[18]   Saxena RK, von Wettberg E, Upadhyaya HD, *et al.* Genetic diversity and demographic history of Cajanus spp. illustrated from genome-wide SNPs. PLoS One 2014; 9(2): e88568.
[http://dx.doi.org/10.1371/journal.pone.0088568] [PMID: 24533111]

[19]   Schierenceck KA, Norman CE. Hybridization and the education of invasiveness in plants and other organisms. Biol Invasions 2009; 11: 1093-105.
[http://dx.doi.org/10.1007/s10530-008-9388-x]

[20]   Nadimpalli RG, Jarret RL, Phatak SC, Kochert G. Phylogenetic relationships of the pigeonpea (Cajanus cajan) based on nuclear restriction fragment length polymorphisms. Genome 1993; 36(2): 216-23.
[http://dx.doi.org/10.1139/g93-030] [PMID: 18469983]

[21]   Songok S, Ferguson M, Muigai AW, Silim S. Genetic diversity in pigeonpea [*Cajanus cajan* (L.)

Millsp.] landraces as revealed by simple sequence repeat markers. Afr J Biotechnol 2010; 9(22): 3231-41.

[22]   Sangiri C, Kaga A, Tomooka N, Vaughan D, Srinives P. Genetic diversity of the mungbean (*Vigna radiata, Leguminosae*) gene pool on the basis of microsatellite analyses. Aust J Bot 2007; 55: 837-47.
[http://dx.doi.org/10.1071/BT07105]

[23]   Tomooka N, Vaughan DA, Moss H, Maxted N. The Asian Vigna genus Vigna subgenus Ceratotropis genetic resources. Dordrecht, Netherlands: Kluwer Academic Publishers 2002; pp. 9-21.

[24]   Al-Saady NAB, Nadaf SK, Al-Lawati AH, Al-Hinai SA. Genetic diversity of indigenous mungbean (*Vigna radiata* L. Wilczek) germplasm collection in Oman. Agric Fores. Fish 2018; 7(6): 113-20.

[25]   Mfeka N, Mulidzi RA, Lewu FB. Growth and yield parameters of three cowpea (*Vigna unguiculata* L. Walp.) lines as affected by planting date and zinc application rate. S Afr J Sci 2019; 115(1/2): 1-8.

[26]   Gondwe TM, Alamu EO, Mdziniso P, Maziya-Dixon B. Cowpea (*Vigna unguiculata* (L.) Walp) for food security: an evaluation of end-user traits of improved varieties in Swaziland. Sci Rep 2019; 9(1): 15991.
[http://dx.doi.org/10.1038/s41598-019-52360-w] [PMID: 31690778]

[27]   Database FAOSTAT. FAOSTAT Database. Food and Agriculture Organisation of the United Nations. Date accessed January 2020. www.fao.org/faostat

[28]   Filippetti A, Ricciardi L. Faba bean: *Vicia faba* L. In: Kalloo G, Bergh BO, Eds. Genetic improvement of vegetable crops. Amsterdam, Netherlands: Elsevier B.V. 1993; pp. 355-85.
[http://dx.doi.org/10.1016/B978-0-08-040826-2.50029-1]

[29]   Ghassemi-Golezani K, Ghanehpoor S, Mohammadi-Nasab D. Effect of water limitation on growth and grain filling of faba bean cultivars. J Food Agric Environ 2009; 7: 442-7.

[30]   Lizarazo CI, Lampi AM, Liu J, Sontag-Strohm T, Piironen V, Stoddard FL. Nutritive quality and protein production from grain legumes in a boreal climate. J Sci Food Agric 2015; 95(10): 2053-64.
[http://dx.doi.org/10.1002/jsfa.6920] [PMID: 25242296]

[31]   Landry EJ, Fuchs SJ, Hu J. Carbohydrate composition of mature and immature faba bean seeds. J Food Compos Anal 2016; 50: 55-60.
[http://dx.doi.org/10.1016/j.jfca.2016.05.010]

[32]   Kindie Y, Bezabih A, Beshir W, *et al.* Field pea (*Pisum sativum* L.) variety development for moisture deficit areas of eastern Amhara, Ethiopia. Hindawi Adv Agric 2019; 1298612: 1-6.

[33]   CSA (Central Statistical agency), Agricultural Sample Survey. 2015.

[34]   Fikere M, Bing DJ, Tedele T, Amsalu A. Comparison of biometrical methods to describe yield stability in field pea (*Pisum sativum* L.) under south eastern Ethiopian conditions. Afr J Agric Res 2014; 9(33): 2574-83.
[http://dx.doi.org/10.5897/AJAR09.602]

[35]   Kosev V, Pachev I. Genetic improvement of field pea (*Pisum sativum* L.) in Bulgaria. Field Veg Crop Res 2010; 47: 403-8.

[36]   Grant JE, Cooper PA, McAra AE, Frew TJ. Transformation of peas (*Pisum sativum* L.) using immature cotyledons. Plant Cell Rep 1995; 15(3-4): 254-8.
[http://dx.doi.org/10.1007/BF00193730] [PMID: 24185786]

[37]   Krejci P, Matuskova P, Hanacek P, Reinohl V, Prochazka S. The transformation of pea (*Pisum sativum* L.): applicable methods of *Agrobacterium tumefaciens*-mediated gene transfer. Acta Physiol Plant 2007; 29: 157-63.
[http://dx.doi.org/10.1007/s11738-006-0020-3]

[38]   Singh S, Gumber RK, Joshi N, Singh K. Introgression from wild *Cicer reticulatum* to cultivated chickpea for productivity and disease resistance. Plant Breed 2005; 124: 477-80.
[http://dx.doi.org/10.1111/j.1439-0523.2005.01146.x]

[39]   Cokkizgin A, Shtaya MJY. Lentil: Origin, cultivation techniques, utilisation and advances in transformation. Agric Sci 2013; 1(1): 55-62.
[http://dx.doi.org/10.12735/as.v1i1p55]

[40]   Rehman S, Altaf CHM. Karyotipic studies in *Lens culinaris* Medic, sup. Macrosperma cv. Laird x Precoz. Pak J Bot 1994; 26(2): 347-52.

[41]   Khazaei H, Caron CT, Fedoruk M, *et al.* Genetic diversity of cultivated lentil (*Lens culinaris* Medik.) and its relation to the world's agro-ecological zones. Front Plant Sci 2016; 7: 1093.
[http://dx.doi.org/10.3389/fpls.2016.01093] [PMID: 27507980]

[42]   Tatsukami Y, Ueda M. *Rhizobial* gibberellin negatively regulates host nodule number. Sci Rep 2016; 6: 27998.
[http://dx.doi.org/10.1038/srep27998] [PMID: 27307029]

[43]   Ferguson BJ. The development and regulation of soybean nodules. In: Board J, Ed. A comprehensive survey of international soybean research- genetics, physiology, agronomy and nitrogen relationship. London, United Kingdom: IntechOpen 2013; pp. 31-47.

[44]   Zhao T, Aleem M, Sharmin RA. Adaptation to water stress in soybean: morphological to genetics. In: Andjelkovic V, Ed. Plant, abiotic stress and responses to climate change. London, United Kingdom: IntechOpen 2018; pp. 33-68.
[http://dx.doi.org/10.5772/intechopen.72229]

[45]   Brar GS, Carter TE. Soybean: *Glycine max* (L.) Merrill. In: Kalloo G, Bergh BO, Eds. Genetic improvement of vegetable crops. Amsterdam, Netherlands: Elsevier V.B. 1993; pp. 427-63.
[http://dx.doi.org/10.1016/B978-0-08-040826-2.50034-5]

[46]   Upadhyaya HD, Dwivedi SL, Ambrose M, *et al.* Legume genetic resources: management, diversity assessment, and utilization in crop improvement. Euphytica 2011; 180: 27-47.
[http://dx.doi.org/10.1007/s10681-011-0449-3]

[47]   Stagnari F, Maggio A, Galieni A, Pisante M. Multiple benefits of legumes for agriculture sustainability: an overview. Chem Biol Tech Agric 2017; 4: 2196-5641.
[http://dx.doi.org/10.1186/s40538-016-0085-1]

[48]   Leon-Lobos P, Way M, Aranda PD, Lima-Junior M. The role of *ex situ* seed banks in the conservation of plant diversity and in ecological restoration in Latin America. Plant Ecol Divers 2012; 1-14.
[http://dx.doi.org/10.1080/17550874.2012.713402]

[49]   Stagnari F, Maggio A, Galieni A, Pisante M. Multiple benefits of legumes for agriculture sustainability: An overview. Chem Biol Technol Agric 2017; 4(2): 1-13.
[http://dx.doi.org/10.1186/s40538-016-0085-1]

Determination of Drought Stress Tolerance Using Morphological and Physiological Markers in Soybean (*Glycine max* L.)

Paseka Tritieth Mabulwana[*] and **Phatlane William Mokwala**

Department of Biodiversity, School of Molecular and Life Sciences, Faculty of Science and Agriculture, University of Limpopo, Private Bag X1106, Sovenga 0727, Republic of South Africa

Abstract: Soybean (*Glycine max* L.) is one of the most important leguminous crop plants worldwide. A lot of attention has been focused on soybean cultivation in South Africa. Its production is affected by several biotic and abiotic stress factors which reduce the yield and quality of the crop. The aim of this study was to evaluate drought tolerance in South African soybean cultivars that have the potential for cultivation in areas where water is a limited resource. Six South African (LS677, LS678, Mopani, Sonop, Knap and Pan1564) and two American (R01416 and R01581) cultivars were carefully studied for morphological and physiological markers using a greenhouse-based study in a randomised block design. The results showed that several morphological (stem length, leaf area, flowers, and seeds) and physiological (chlorophyll, moisture content, phenolics, flavonoids, ureide content and antioxidant activity) parameters were affected by drought stress. However, cultivars with high phenolic and flavonoids content were associated with high antioxidant activity and slightly increased their yields than other varieties. The anatomical analysis also showed some interesting differences in response to reduced water treatment, with the sizes of vascular tissues and sclerenchyma tissues decreasing under drought stress. In conclusion, this study indicated that reduced stem length and inability to reduce leaf area by soybean plants could lower plant growth and yield response under drought stress. In addition, increased chlorophyll and secondary metabolite content can improve soybean growth under limited water conditions.

Keywords: Anatomy, Antioxidant activity, Chlorophyll, Morphology, Phenolics, Soybean, Total flavonoids.

INTRODUCTION

Glycine max L. is considered to be one of the most important grain legumes in sub-Saharan Africa and the world at large.

---
[*] **Corresponding author Paseka Tritieth Mabulwana**: Department of Biodiversity, School of Molecular and Life Sciences, Faculty of Science and Agriculture, University of Limpopo, Private Bag X1106, Sovenga 0727, Republic of South Africa; Tel: +2715 268 3344; E-mail: Paseka.Mabulwana@ul.ac.za

Phetole Mangena (Ed.)

The cultivation of this crop has also been going on for decades in many parts of Africa, particularly Nigeria and South Africa, where it amounts to over 35% of the total grain legume production in this region [1]. Soybean remains one of the less cultivated food crops in Africa compared to maize, cassava, and sorghum. In South Africa, the cultivation of this crop is widespread. According to earlier reports by the Department of Agriculture, Forestry and Fisheries; major areas of cultivation per province are the Mpumalanga highveld, Free State highveld, areas around Pietermaritzburg in the KwaZulu Natal province, Northwest province, Gauteng, and Limpopo. However, soybean production in the Southern part of South Africa, such as the Northern, Eastern and Western Cape remains very minimal due to cultivation conditions and unpredictable weather patterns. Furthermore, the chief soybean producers in Africa (Nigeria and South Africa), as well as other countries like Zambia, Tanzania and Zimbabwe, have also slightly increased their statistical production records in 2010 and beyond, whilst cultivation continues to show fluctuating results [1, 2].

The cultivation problems emanate from the various climatic challenges that largely affect crop growth and productivity, such as reduced seed viability, pod shattering and lack of genetically improved cultivars, especially those conferring tolerance to both biotic and abiotic stress factors. Generally, crop plant growth and yield become severely impacted by abiotic stress, such as the inadequate supply of water, resulting in decreased carbon assimilates contents, increased susceptibility to insect pests and diseases. Evidence of the occurrence of fungi and insect-induced stalk rots, wilts and foliar diseases caused by water deficit stress were also reported in soybean plants [2, 3]. As indicated by Singh and Singh [4], abiotic stress perpetuates challenges facing the growth and productivity of soybean, particularly, in developing countries where cultivation remains stagnated and very limited due to the increasing populations. As such, these impact negatively on the wellbeing of many people, and their gradually developing economies.

Additionally, these also pose more problems to the quantity and quality of processed soy-based products that are being locally manufactured. Some of those manufactured products include soy flour, milk, textured soy protein and oil. Soybean is also regarded as one of the most important sources of affordable protein and vegetable oil compared to wheat and maize [5]. Apart from being used as indicated above, it is also used as a source of biodiesel and animal feed [6]. This crop is also useful in the improvement of soil fertility and the capability to take fixed atmospheric nitrogen [2, 3]. Nitrogen fixation is brought about by a mutualistic relationship between soybean and *Bradyrhizobium* bacteria, which forms nodules in the roots. The microorganisms help the plant in fixing or converting nitrogen into forms usable by the plant-like NH_3, NO_3^- and NH_4^+.

Among the major abiotic stressors, drought is one of the most damaging constraints that adversely affect soybeans in many critical aspects of their growth and metabolism. Soybean's growth, development and productivity are severely diminished when soil and cell water potential becomes inadequate to sustain metabolic functioning. However, little has been done to gather adequate comprehensive information regarding the specific changes that take place in water-stressed plants at the anatomical, physiological, and morphological level, especially for soybean varieties cultivated in Africa [2]. Therefore, this chapter discusses the determination of drought stress tolerance using various morphological and physiological markers, compared between several soybean cultivars (Mopani, Sonop, Knap, Pan, LS677, LS678, R01416 and R01581). The effects of water deficit stress were assessed at moderate and severe stress levels, and results obtained were compared to the controls, which were watered daily to saturation depending on the growth medium moisture content.

STRATEGIES TO IMPROVE CROP PRODUCTIVITY UNDER DRY LAND AND RAIN-FED CULTIVATION

According to Liu *et al.* [7], dryland comprises about 40% of the global land area. This estimates include land which receives more than 2,000 mm rainfall as dryland. Despite the broad definition of dryland, in general, lands that receive very limited precipitation are regarded as dryland. Agricultural development in drylands, therefore, involves intensive labour and financial resources, but consequently, these investments are usually accompanied by highly reduced large scale total crop yields as a result of the variability and instability of the environment [8]. However, large scale commercial crop cultivation relies upon irrigation to improve the yield of the crops compared to dryland or rain-fed farming. Although irrigation improves yield to about two times that of rain-fed farming as indicated by Babovic and Milic [8], the globe cannot afford increased irrigation due to the experienced depleted water resources as a result of severe drought in many parts of the world.

Climate change, especially rainfall patterns, always determines the water availability, evaporation rates and the effectiveness of irrigation in agriculture. Effective irrigation could be achieved when there is adequate precipitation to sustain crop growth and increase yield. Future crop improvement strategies, farming practices and farm management should be based on an interdisciplinary approach when dealing with causative agents of environmental instability. To achieve this goal, inputs are needed from environmental scientists, breeders, agriculturists, agricultural extension specialists and meteorologists working closely together to strive for improvement and stability of crop performance and production, especially under rain-fed farming conditions [9]. Some of the cultivation practices that can be optimised and employed to improve crop production in dryland farming include the following:

Double or Intercropping

Double cropping refers to a sustainable agricultural practice in which more than one crop is grown on the same ground during the same period of time. Irrigation helps increase and stabilise crop production and also helps to promote this cropping system. Growing different crops in close proximity could help promote water retention capacity and soil quality. According to Babovic and Milic [8], this system has the advantage of increasing crop and land productivity, which help in boosting the economy, feeding the ever-increasing population and alleviating the effects of hunger.

Tillage

Tillage simply means agricultural preparation of the soil for growing crops. This is the preparation of soil by mechanical agitation of various types, such as digging, stirring, and overturning. This process may leave plant material on the ground to decompose, manipulating optimised soil nutrient conditions for successful plant establishment and enhancing the growth of the desired crops. Tillage practices are very effective in reducing soil erosion by wind or water, runoff, and water evaporation. This is achieved by promoting water infiltration and moisture retention. This practice promotes efficient water use and conservation as it allows the soil to absorb more water by precipitation or irrigation, while losing very little water through the process of evaporation [10]. Although tillage practices also have some reported disadvantages, this process could still be effectively appropriate and, therefore, used by the agricultural industry to improve food crop production, especially in soybean [11].

Conventional Plant Breeding

Conventional plant breeding involves changing the genetic composition to improve crop varieties. The cultivars are improved for tolerance to environmental stress factors (drought, salinity, high temperature *etc.*), diseases (pests) and also to improve yield and quality of seeds, as well as the nutritional composition. The traditional methods involve crossing two plants (male and female) to combine the desired traits from both parents [12]. Conventional breeding is regarded as an extremely important tool, but, it also has some limitations. A cross between two parents may result in the progeny inheriting a mixture of genes (both desirable and negative traits). As a result of this mixture of genes, plant breeders may need to back cross the progeny, which is a labor intensive exercise, time-consuming and also requires sophisticated equipment and techniques [12].

Plant Biotechnology

Genetic transformation is one of the approaches currently used for the improvement of many cereal and grain food crops. The practice involves the ability to insert foreign DNA segments of interest into host crop species, aimed at altering the host genetic makeup. For soybean, this technique was first used in the late 1980s, through the particle bombardment (biolistic) method and *Agrobacterium tumefaciens* or *A. rhizogenes* based protocols. The use of single-celled bacterium like those mentioned above is the most preferred for genetic manipulation of plants. *Agrobacterium*-mediated genetic transformation has been used to develop soybean cultivars tolerant to agrochemicals, pathogens, and pests. The examples include Roundup Ready genetically modified (GM) soybeans that currently dominate the market, accounting for more than 83%, 94%, and 100% of total production in the United States, Brazil, and Argentina, respectively [13].

However, the use of *Agrobacterium* in soybean improvement for tolerance and resistance against abiotic stress, such as water deficit, high temperatures and salinity, especially for cultivation in drier areas still remains a massive challenge. But genetic engineering has emerged as a very successful breeding tool for breeders and the agricultural community at large. Plant genetic transformation differs from conventional breeding mainly because instead of mixing a lot of genes from the sexual parents, only a desired specific gene will be isolated and inserted into a plant of interest. This plant breeding technique is convenient, rapid, and affordable for many laboratories. The lant transformation technique can be used to create new genetic combinations to improve the genetic diversity of soybean, since it could also allow for the expression of genes from unrelated sources. This proves that genetic transformation remains one of the imperative tools required to solve global environmental challenges [14].

Similarly, conventional breeding and biotechnology require prior knowledge of cultivars or varieties that have consistently presented desirable morphological and reproductive traits [15]. Furthermore, they both require compulsory screening and selection of suitable traits as some of the very critical steps preceding breeding. Therefore, both techniques could be used in counteracting the negative impact of the shortage of rainfall and assist in the development of varieties showing tolerance to drought in order to improve agricultural outputs or yield.

ADAPTATION OF PLANTS TO DROUGHT STRESS

Drought stress is a very complex process that negatively affects the growth and productivity of soybean plants. Soybeans respond to water deficit stress differently. However, the type of mechanism used by plants in general to adapt to dry conditions depends upon the type of species, growth habit, growth stage during which stress is imposed, and the intensity of stress [15, 16]. Plants possess several adaptive traits to endure the period of stress. Some of the devised mechanisms for drought stress endurance include escape, tolerance, and avoidance. They escape drought stress by shortening their life cycle, maturing earlier for flowering, fruiting, and seed dispersal. Drought stress avoidance meanwhile involves reduced water loss through leaves and increased water absorption by the roots. Drought tolerance is, however, a very complex mechanism that involves several aspects that include osmotic adjustment and increased antioxidant activity. During this phenomenon, plants use a series of morphological, physiological, cellular, and molecular processes to respond to the imposed drought stress effects [15, 17].

Morphological Adaptations

Heschel and Riginos [18] reported that plants respond to water stress by reducing their leaf sizes to maintain high water potential in the cells. A typical example of such plants is geophytes that survive drought stress by losing all of their vegetative parts during the period of water deficit stress and then rise again when water becomes available [19]. The morphological features associated with drought stress in plants include but are not limited to reduced plant height, leaf surface area, as well as flowers and pod abortion [20]. In this study, water limitations caused a decrease in stem length, from the tallest stems observed in the control (1500 mL/pot) to shorter stems following moderate (300 mL/pot) and severe drought stress treatment that received less water (about 150 mL per day/pot). Stem length was decreased by 5% in the moderate water treatment and to more than 15% in severe water-stressed plants. These effects were more pronounced in LS678 and Pan1564 genotypes than R01416F and R01581F soybean genotypes (Table **1**).

The findings demonstrated that drought caused negative effects on plant height, as indicated by the six South African cultivars used, especially in contrast to the American varieties. Such effects were also reported by Desclaux *et al.* [21], using cultivars of both determinate and indeterminate growth at 30 and 50% water stress under greenhouse conditions. In contrast, better and improved plant heights were recorded in soybean cultivar R01416F and R01581F, as also reported by Chen *et al.* [22]. The American cultivars were included in this study as positive controls since they were reported and registered for improved yield and nitrogen fixation under drought stress [22]. Reduced stem height may have direct effects on the number of flower buds formed per plant. In general, plants with increased stem height produced a high number of flowers compared to those with shorter stems, and this may have a direct impact on the quantity of yield harvest.

Table 1. Mean stem length (cm) of the cultivars from different treatments determined at R3 growth stage after plant exposure to drought stress.

Soybean Genotype	Moderate Stress	Severe Stress	Well-Watered Control
Pan1564	22.1 ± 0.14[f]	21.6 ± 0.77[h]	23.4 ± 0.28[h]
Knap	37.7 ± 0.56[b]	36.6 ± 0.14[a]	39.2 ± 0.42[c]
Mopani	32.1 ± 0.21[c]	25.8 ± 0.14[f]	33.2 ± 0.35[d]
Sonop	37.2 ± 0.07[b]	35.3 ± 0.42[b]	40.3 ± 0.28[b]
LS677	31.6 ± 0.28[c]	28.3 ± 0.14[d]	32.5 ± 0.28[e]
LS678	39.2 ± 0.28[a]	27.3 ± 0.07[e]	50.1 ± 0.56[a]
R01581F	30.3 ± 0.21[d]	31.4 ± 0.07[c]	31.2 ± 0.84[f]
R01416F	23.4 ± 0.49[e]	23.2 ± 0.21[g]	26.4 ± 0.35[g]

Values are means comparison of stem lengths ± standard deviation. Means with different letters within columns are statistically different at 5% confidence level according to t-test.

According to Table **2**, the number of flowers, leaf surface area and relative leaf water content were significantly reduced by exposing soybean plants to water deficit stress, particularly in cultivar Pan1564, Sonop, Mopani, Knap, LS677 and LS678, respectively. The observed findings also indicated that flowering was also slightly and negatively impacted in the two cultivars used as positive controls (Table **2**). The highest mean number of flowers was produced by cultivar R01416F, which also recorded the highest mean leaf surface area (44 cm^2). A different trend was observed in all other seven cultivars, which responded to drought stress by decreasing their leaf surface area. Furthermore, drought stress also caused significant reductions in the relative leaf water content of soybean R01581F, Knap, Sonop, Mopani and Pan1564 compared to LS677, LS678 and R01416F. However, R01416F gave the largest relative leaf water content compared to the LS soybean cultivars (Table **2**).

Table 2. Mean number of flowers, leaf surface area and relative leaf water content measured at R3 growth stage after plant exposure to drought stress.

Soybean Cultivar	Flower Number	Leaf Surface Area (cm^2)	Relative Leaf Water Content (%)
Control			
Pan1564	10 ± 4.24[a]	46 ± 3.88[d]	94 ± 2.57[a]
Knap	10 ± 0.00[a]	42 ± 0.36[f]	83 ± 2.16[c]
Mopani	9 ± 2.82[b]	114 ± 2.14[a]	77 ± 0.76[f]
Sonop	9 ± 0.70[b]	49 ± 0.56[c]	90 ± 0.73[b]
LS677	8 ± 2.12[c]	45 ± 3.14[c]	94 ± 0.24[a]

(Table 2) cont.....

Soybean Cultivar	Flower Number	Leaf Surface Area (cm²)	Relative Leaf Water Content (%)
LS678	10 ± 2.12^a	38 ± 1.36^g	88 ± 1.13^d
R01581F	8 ± 1.41^c	58 ± 1.57^b	88 ± 1.33^d
R01416F	10 ± 1.41^a	28 ± 1.64^h	89 ± 1.49^c
Moderate Stress			
Pan1564	9 ± 3.53^b	29 ± 2.81^f	83 ± 1.72^d
Knap	8 ± 3.53^c	29 ± 1.53^f	69 ± 1.57^h
Mopani	8 ± 0.70^c	66 ± 0.81^a	74 ± 1.52^g
Sonop	6 ± 5.65^e	38 ± 3.81^d	84 ± 1.43^c
LS677	8 ± 0.70^c	36 ± 0.72^e	80 ± 1.36^f
LS678	8 ± 0.00^c	25 ± 0.17^g	88 ± 0.32^b
R01581F	10 ± 1.41^a	40 ± 1.68^c	82 ± 0.92^e
R01416F	7 ± 0.70^d	44 ± 0.71^b	89 ± 0.65^a
Severe Stress			
Pan1564	7 ± 1.41^b	10 ± 1.44^g	65 ± 0.89^e
Knap	6 ± 0.00^c	21 ± 0.00^c	60 ± 0.42^g
Mopani	6 ± 2.82^c	39 ± 1.57^a	61 ± 0.71^f
Sonop	5 ± 1.41^d	19 ± 2.40^d	80 ± 2.14^b
LS677	6 ± 0.00^c	21 ± 0.19^c	88 ± 1.47^a
LS678	7 ± 4.24^b	16 ± 3.98^e	75 ± 2.81^d
R01581F	8 ± 0.70^a	15 ± 0.92^f	77 ± 0.72^c
R01416F	8 ± 0.70^a	24 ± 0.63^b	75 ± 0.68^d

Mean values with different letters are statistically different at 5% confidence level according to t-test.

The overall reduction in plant traits, including those discussed above, may lead to complete suppression of the growth and distribution of roots, stems and leaves in water-stressed plants. Normally, when vegetative shoot growth gets diminished by the induced stress, flower abortion may occur, which subsequently leads to the total halt and reduction of the plants' capacity to produce seeds. As the level of drought stress becomes intense, the level of water in the growth medium is also decreased, consequently affecting the rate of metabolic activities. Under the given artificial conditions, the limited water supply decreased moisture content of the vermiculite (growth medium) and more losses were recorded due to the prolonged period of water deficit. Moreover, the plants also continued to lose more water through the process of transpiration [23]. As generally observed, the ground moisture content is greatly decreased and it becomes insufficient for the plant to carry out its daily life metabolic functions, particularly under dry seasons. In this case, cell water potential becomes inadequate for sustaining metabolic functioning and thus, plant growth, development and productivity are severely diminished [24].

Physiological Adaptations

Plants can escape drought stress by rapidly increasing their growth rate and reach their maturity stages before the stress becomes intense. They can also adapt to water stress physiologically, absorbing more water while reducing the rate of water loss *via* transpiration, as indicated previously. Loss of water *via* transpiration can be decreased by physiologically lowering stomatal conductance and reducing the leaf surface area. This is achieved through osmotic adjustment and maintaining cell turgor pressure during drier periods [25]. The two main mechanisms that are generally targeted for this purpose include non-enzymatic/enzymatic processes and pathways that function to regulate and scavenge reactive oxygen species (ROS), that may damage essential tissues and cell components, as well as interfering with important cellular processes [26]. Non-enzymatic mechanisms employ several secondary metabolites to prevent the formation of reactive oxygen species. While enzymatic pathways involve the use of different enzymes to eliminate unwanted toxins [27].

Reports showed that drought stress could cause imbalances of many enzymatically regulated processes and alter the natural status of the cell environment, which drastically disrupt processes leading to successful crop growth and yield. It limits photosynthesis by influencing carbon dioxide diffusion, stomatal conductance and carboxylation/reduction processes taking place in the palisade mesophyll tissues of leaves. As indicated by Pinheiro and Chaves [28], the photosynthetic activity will be decreased due to increased chlorophyll bleaching, reduced CO_2 diffusion and the concomitant formation of ROS. Basically, the formation of these deleterious effects will take place in soybeans undergoing water stress than well-watered plants. All these effects illustrate why water remains a critical component in plant cells, including many cellular processes like the production of organic metabolites, hormones biosynthesis and transport of organic and inorganic acids.

Chlorophyll Content

As also observed in this study, water deficit had a negative effect on percentage chlorophyll content emanating from the relative leaf water content. The results were indicated that, total chlorophyll content was reduced by water levels, from moderate to severe induced drought stress (Fig. **1**). This appeared severely true for all studied soybean genotypes except LS677, R01416F and R01581F where the tissue chlorophyll content slightly fluctuated among the treatments. Soybeans, R01416F and R01581F, showed a strong fluctuating trend, and this might have been due to their slight tolerance of water deficit stress. The chlorophyll content is a key factor, which affects the rate of photosynthesis, therefore, controlling the growth rate of plants [29]. As expected, the chlorophyll content was higher in well-watered plants compared to plants subjected to water-stress treatments. Hassanzadeh *et al*. [30], cited the destruction of chlorophyll molecules by lipid peroxidation triggered by reactive oxygen species' generation during water-deficit stress for this kind of response.

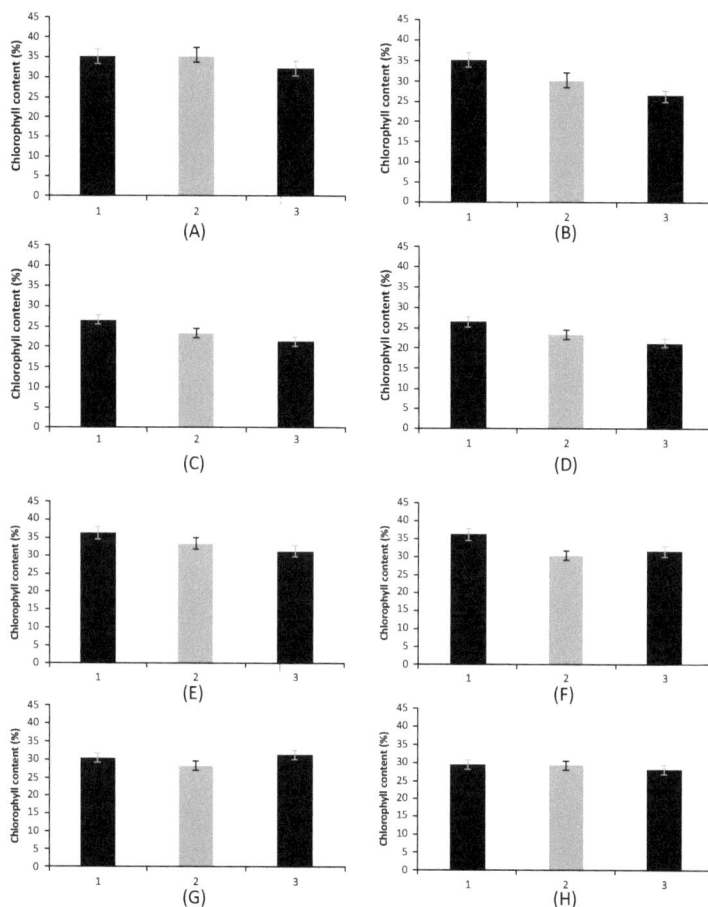

Fig. (1). Overall mean percentages of chlorophyll content measured following eight weeks (8) of water deficit stress. Total chlorophyll content per soybean cultivar are referred to as follows; Mopani **(A)**, Sonop **(B)**, Knap **(C)**, Pan1564 **(D)**, LS678 **(E)**, LS677 **(F)**, R01416F and R01581. Axis title 1, 2 and 3 refers to the control, moderate and severe stress levels, respectively.

In their study, Hassanzadeh *et al*. [30], highlighted varied metabolite responses based on genetic integrity, where sesame genotypes yielded more chlorophyll content than those that produced lower amounts of these pigments when grown under drought stress conditions. These findings imply that the total chlorophyll content may be used as a stress indicator to distinguish between stress-sensitive and stress-tolerant soybean cultivars. Reduced chlorophyll content in the leaves of soybean subjected to water deficit stress was also recently reported by Masoumi *et al.* [26], as a morpho-physiological marker for drought stress tolerance.

Production of Antioxidants (Non-Enzymatic Mechanism)

Antioxidants refer to secondary metabolites produced by plants when undergoing physiological water-deficit stress. These compounds are mainly produced as a form of a defense mechanism against oxidative stress caused by free radicals or against herbivory. Such organic compounds include flavonoids, hydrolysed /condensed tannins, phenolics, ascorbic acid and glutathione (Fig. **2**) [31]. Many food products such as fruits, vegetables and grains contain antioxidants. These compounds are capable of delaying, retarding, or minimising the development of rancidity, thus maintaining nutritional quality and increasing the product storage quality. Like other plants, soybeans use antioxidant systems to scavenge or prevent the formation and deleterious effects of reactive oxygen species. The commonly expressed secondary metabolites serving as antioxidants in soybean as a result of drought stress include ascorbic acid and phenolic compounds [32 - 34].

Soybean Total Phenolics and Total Flavonoids

Phytochemicals such as flavonoids, phenolics and tannins have gained considerable interest among consumers and scientists [35]. Soybeans, including plants such as watermelon (*Citrullus lanatus*) and gooseberries (*Ribes grossularia*) proved to contain large amounts of these phytochemicals and, therefore, serving as a decent source of antioxidants. Reports have indicated that plant tissues contain generally higher total phenolics than the flavonoid content. In this study, a spectrophotometric determination of phenolics and flavonoids was performed according to the Folin-Ciocalteau method using Gallic acid standards [36] and the Catechin method as described by Marinova *et al.* [37], respectively. Soybean LS677, LS678 and R01416F also demonstrated a higher phenolic content than R01581F, Mopani and Sonop. But, cultivar Mopani, Sonop and R01581F produced higher concentrations of phenolics under moderate drought stress than control condition and severe drought stress (Table **3**).

There were high variations observed in phenolic expression in the treatment of all genotypes. With regard to this, Knap, LS677, R01416F and R01581F showed high levels of total phenolics when subjected to moderate stress and during the period of no water-deficit stress. In comparison, Mopani and Sonop produced the highest phenolic content under severe water stress, with Mopani expressing more phenolics than Sonop under this treatment (Table **3**). These varied patterns of responses demonstrate the sensitivity and susceptibility of each genotype towards the physiological status of the plant, as influenced by drought stress. Furthermore, according to the results, soybean Pan1564 showed the highest concentration of flavonoids under severe water stress. For Knap, the total flavonoids were significantly expressed and high in all plants exposed to water-deficit stress.

In contrast to Knap, Mopani showed a higher concentration of flavonoids than Sonop but, the flavonoid concentration decreased as water levels were depleted (Table **3**). The concentration of flavonoid content continued to vary according to genotype and water stress conditions. These findings further implied that the increased expression of total flavonoids, as well as other phenolics in plant tissue may be indeed linked to drought stress tolerance.

Antioxidant Activity of Selected Soybean Genotypes

Soybean genotypes were also reported by Peiretti *et al.* [38], to contain variably high and different amounts of phytochemicals and antioxidants. Among these, some of the major sources of phytochemicals include green leafy vegetables and red-flashed fruits, which also serve as a rich source of energy, proteins, essential oils, and micro/macro-nutrients [35]. There are many phenolic compounds produced in plants for diverse functions. The different compound types include but are not limited to isoflavones, benzoic acids, phenolic acids, and flavonoids with a flavone backbone. One of the major functions of these compounds is the strong antioxidant activity and potential to readily eradicate the negative effects of ROS. The antioxidant activity of the analysed phenolics was determined using ascorbic acid standard in the 2,2-Diphenyl-1-picrylhydrazyl (DPPH) (Fig. **2**) method as described by Odhav *et al.* [39].

As indicated in Table **3**, reduced cell water potential in soybean tissues increased the antioxidant activity of Mopani, Klap, LS677 and LS678. Meanwhile, the antioxidant activity of R01416, R01581, Sonop and Pan fluctuated, exhibiting slight differences among the cultivars and the three water regimens used. According to the findings of the study, Sonop demonstrated the highest antioxidant activity of 77.2% DPPH activity, accompanied by a significant amount of flavonoids (16.6 µg/g) and phenolics (28.5 µg/g) under severe drought stress. Soybean cultivars Knap, R01416 and R01581 also showed a high antioxidant activity as reported by Malencic *et al.* [40]. Generally, the cultivars that showed a high concentration of total phenolics also showed high antioxidant activity. Furthermore, water stress increased the antioxidant activities of most of the soybeans, implying that cultivars with such high activity may be adapted to a higher yield than those with lower antioxidant activities under limited water availability.

Table 3. Phytochemical analysis of the soybeans after plant exposure to drought stress.

Soybean Cultivar	Total Phenolics (µg/g Fresh Sample)	Total Flavonoids(µg/g Fresh Sample)	Antioxidant Activity (% of Gallic Acid)
Control			
Pan1564	27.3 ± 0.06^e	20.8 ± 0.88^a	36.2 ± 2.77^e
Knap	24.5 ± 0.10^g	19.9 ± 0.64^b	38.8 ± 0.17^d
Mopani	26.6 ± 0.08^f	19.1 ± 0.14^c	42.6 ± 0.17^c
Sonop	24.9 ± 0.07^g	15.5 ± 0.56^e	75.9 ± 0.24^a
LS677	30.7 ± 0.02^c	19.7 ± 0.14^b	30.9 ± 0.40^f
LS678	33.1 ± 0.02^b	18.5 ± 1.36^d	75.5 ± 0.68^a
R01581F	29.6 ± 0.03^d	19.4 ± 0.57^b	47.8 ± 0.32^b
R01416F	65.1 ± 0.00^a	19.3 ± 0.04^c	47.6 ± 1.29^b
Moderate Stress			
Pan1564	38.6 ± 0.03^e	16.4 ± 0.81^d	47.5 ± 0.77^d
Knap	46.3 ± 0.04^d	20.7 ± 0.03^a	49.9 ± 0.09^c
Mopani	24.7 ± 0.14^g	18.7 ± 0.22^c	44.3 ± 0.61^f
Sonop	24.7 ± 0.02^g	14.8 ± 0.81^e	70.4 ± 1.62^a
LS677	48.5 ± 0.04^c	19.4 ± 0.72^b	64.1 ± 0.86^b
LS678	64.0 ± 0.00^a	19.7 ± 0.17^b	35.7 ± 0.75^g
R01581F	25.3 ± 0.02^f	17.9 ± 0.68^d	42.1 ± 0.10^f
R01416F	59.7 ± 0.00^b	18.4 ± 0.01^c	45.6 ± 0.05^e
Severe Stress			
Pan1564	35.7 ± 0.05^d	14.1 ± 0.44^f	46.9 ± 0.92^f
Knap	30.5 ± 0.06^g	20.2 ± 0.00^a	53.3 ± 0.23^d
Mopani	32.9 ± 0.03^f	14.6 ± 0.57^f	45.5 ± 0.49^g
Sonop	28.5 ± 0.11^d	16.6 ± 1.40^e	77.2 ± 1.15^a
LS677	56.2 ± 0.02^a	18.5 ± 0.11^c	75.5 ± 0.68^b
LS678	52.3 ± 0.22^b	19.0 ± 3.98^b	61.7 ± 1.50^c
R01581F	41.4 ± 0.00^c	17.2 ± 0.02^d	46.3 ± 0.31^f
R01416F	33.5 ± 0.07^e	17.9 ± 0.63^d	49.1 ± 0.89^e

Mean values with different letters within columns are statistically different at 5% confidence level according to t-test.

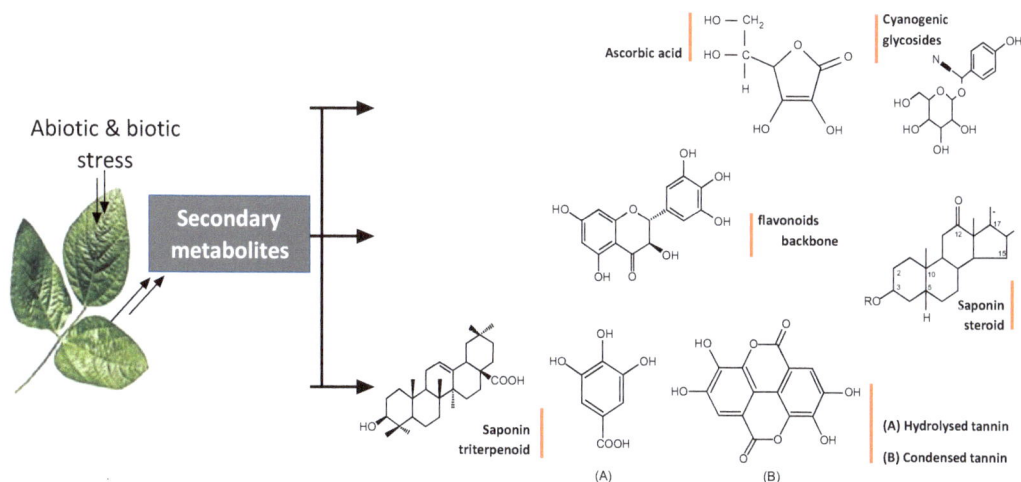

Fig. (2). Basic structure of some of the secondary metabolites produced as a result of plant sensitivity and altered physiological processes due to changes in cell water potential under various growing conditions or caused by other types biotic or abiotic stress.

Enzymatic Radical Scavenging Mechanisms

The enzymes involved in scavenging ROS include superoxide dismutase, catalase and glutathione peroxidase. These enzymes play a key role in helping the plants to quench or prevent the formation of toxic compounds in the cells to minimise oxidative stress damage [41]. The enzymes catalyse the conversion of ROS to less harmful substances such as hydroxide ions, nitrous acid, and water [42]. For example, superoxide dismutase is a key enzyme that catalyses the dismutation of superoxide into oxygen and hydrogen peroxide, meanwhile catalases are responsible for catalysing the conversion of hydrogen peroxide into oxygen and water. The latter is very important because of its highest turnover number, which is the ability to carry out millions of reactions in a second. The glutathione peroxide is involved in the reduction of lipid hydrogen peroxide to alcohol and it is also involved in the reduction of free hydrogen peroxide to water that takes place in the cell cytoplasm [26].

Ureides Production

Nitrogen fixation rapidly declines when legumes are exposed to water deficit stress. The metabolic products of nitrogen fixation in legumes include ureides that accumulate in plant tissues as a result of water deficit stress. Various reports have indicated that this accumulation has been strongly correlated with the inhibition of nitrogen fixation that takes place in legumes [43]. Alamillo *et al.* [43], reported ureide accumulation under drought stress using molecular analysis of allantoate in roots, shoots, leaves and nodules of *Phaseolus vulgaris*. Findings showed that the expression of the ureide metabolism genes analysed was also induced by abscisic acid (ABA), suggesting the involvement of this plant hormone in ureide accumulation. Ureides were reported by Zrenner *et al.* [44], to be formed in the plant fraction of the nodules through a *de novo* purine biosynthesis and purine oxidation, where xanthine dehydrogenase and urate oxidase served as key enzymes in this ureide biosynthesis.

In this study, ureides were extracted from the frozen leaf and nodule materials in liquid nitrogen using a protocol described by Van Heerden *et al.* [45]. All cultivars used in this study (Pan1564, Mopani, LS677, LS678, R01416F and R01581) indicated a higher concentration of ureides (allantoin and allantoic acid) in the leaves subjected to moderate water stress. However, cultivar Knap has reacted slightly differently with ureides accumulation increasing with the decrease in water availability. This affected the transport of ureides from the source to other parts of the plant, where these metabolites will be used as a fixed form of nitrogen. The study showed evidence of drought stress interference with metabolic activities of nodules, leading to ureides accumulation that may cause inhibition of feedback of the nitrogen fixation process [46]. In this study, the concentration of ureides was determined from the leaves at the R3 growth stage.

Most soybeans showed higher ureides' concentration under moderate stress treatment followed by severe water deficit; meanwhile, well-watered plants had the lowest amounts. Typically, ureides concentrations increased with increasing water deficit stress. Thus, these were clear indications that moderate stress rather than severe water deficit is somewhat suitable for nitrogen fixation. In the control, which was well-watered, the amount of water supply could have been too high with limited oxygen availability for the roots to carry out sufficient fixing of nitrogen. Landrera *et al.* [47], also reported similar findings in which nitrogen fixation was inhibited in soybean cultivar "Bixoli" which showed great sensitivity to drought stress.

HISTOLOGICAL EXAMINATION OF LEAVES, PETIOLES, ROOTS AND NODULES

To assess cell structures or characteristics of the leaves, petioles, roots, and nodules of water-stressed plants, histological examination through microscopy was performed. Several histological techniques are commonly used for examining plant tissues, especially to gather and provide gross essential information that may not be evident upon a mere visual inspection. Microscopic techniques were used in this study to scrutinize anatomical features of the roots, leaves and nodules to examine the nature and form of cell division during moderate and severe water-deficit stress (Fig. **3**). The anatomical structures of the eight soybean cultivars in all three water treatments were studied based on the transverse sections of the selected parts and organs, as described by Rajan [48].

Histological Analysis in Roots

In general, roots have a vascular system at the centre called a compacted central stele. Protoxylem and phloem in dicots are always arranged in an alternating manner around this stele (Fig. **3B**). Metaxylem appears to be located on the inner side of the protoxylem, as indicated in the figure. Covering the stele is a thick layer of endodermal cells, and the cortex is found outside the stele and is composed of the ground tissue, mainly the parenchyma cells. An extensive xylem tissue was uniformly produced at the centre of the root, which also seems to have undergone secondary growth, pushing the cortex to the outside. The formation of extensive xylem tissues at the central stele in all soybean cultivars could be an adaptation for growth under limited water conditions.

Similar observations were previously reported by Prince *et al.* [49], when evaluating root xylem plasticity to improve water use and yield in water-stressed soybeans, and earlier by Russin and Evert [50] in *Populus deltoides*, a dicotyledonae cottonwood of the family Salicaceae. According to Prince *et al.* [49], increased metaxylem number in the functions of the root as an adaptation mechanism to drought, improved root hydraulic conductivity by reducing the metabolic cost of exploring water in deeper soil strata and enhanced the transport of readily available water.

Leaf and Petiole Anatomy

Like other dicotyledonous plants, the anatomy of all soybean leaf tissues was composed of both the spongy and palisade mesophyll, covered by the adaxial (upper) and abaxial (lower) epidermis (Fig. **3C-F**). The upper dermal cells were thicker, with heavier cuticle than the lower epidermis which had a high density of stomata. A mid-rib (midvein) comprising a big vascular bundle was also visible at the centre of the leaf transverse section. The midvein of the cultivar R01581F treated with moderate water-deficit stress displayed rather medium-sized xylem vessels located adjacent to the sclerenchymatous bundle sheaths (Fig. **3D**) compared to all cultivars used. However, LS677 presented unusually parenchymatous fibre cells surrounding the midvein under similar water stress conditions, with the thicker layers of both the upper and lower epidermal cells. Furthermore, LS677 plants subjected to severe drought showed a significant variation in the size of xylem vessels in the midvein, with far smaller protoxylem vessels than those observed under moderate stress.

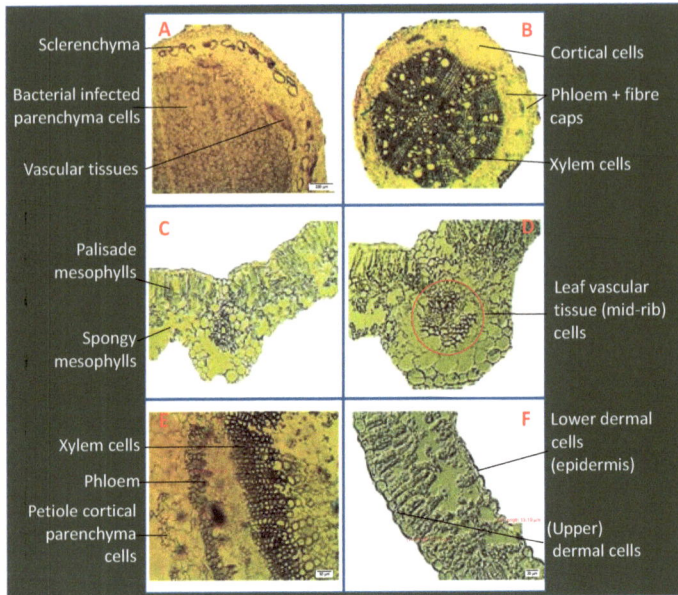

Fig. (3). Examples of transverse sections of soybean plants subjected to water deficit stress. Soybean nodule anatomy of the cultivar Mopani under severe drought stress (**A**). R01581F root cross section of well-watered plant (**B**). The cross section of the leaf anatomy of the cultivar R01581F (**C**), mid-vein structure (**D**), leaf petiole (**E**) and illustration of the lower/upper dermal cells (**F**).

Such observations were also made in the leaf stalk (petiole) under similar water stress conditions. Although, unlike the anatomy of the leaf and root, the pith occupied the centre of the petiole as observed in the stem cell structure. But, the vascular bundles were not clearly defined and arranged as observed in the stems. The phloem tissue was heavily surrounded by layers of fibre caps adding strength to the tissues, probably an adaptation for drought stress (Fig. **3 E**). It was not clear whether the number of xylem vessels would enhance the efficiency of water uptake in leaves during water-limited conditions [49]. The varied anatomical characteristics exhibited by these cultivars in their leaves and petioles indicated that the soybeans modified their plant tissue architecture to achieve adaptation to a water-limited environment. This was also reported by Den Herder *et al.* [51], elucidating a modified root system architecture as a proposed pipeline for enhanced crop growth and yield.

Nodulation

The formation of cell protuberance containing nitrogen-fixing Gram-negative bacteria in the roots of legumes plays a critical role in improving plant growth and yield [24]. Nitrogenase enzymes are responsible for fixing atmospheric nitrogen and this requires an anaerobic environment in nodules protected by a layer of sclerenchymatous tissues that help to block oxygen from entering the nodule cells. The establishment of nodules guarantees the supply of fixed nitrogen for the production of proteins, nucleic acids and other nitrogen-containing metabolites in the plant [24]. This study has observed a clear variation in the anatomy of nodules produced by the soybean plants under different water treatments. The cultivar R01581F produced the most intense sclerenchymatous layer of cells surrounding the nodules, and there were no noticeable changes detected in other soybean cultivars used (Fig. **3A**).

The layer of sclerenchyma in the root nodules of R01581F was large and thickened with fewer infected parenchyma cells and one large vascular bundles. This anatomical architect may have been conducive to nitrogen fixation under water deficit stress. On the other hand, the large nodule vascular systems facilitated the transportation of available water, dissolved minerals and food substances throughout the root nodules. Furthermore, nodulation, as well as, nodule anatomy were highly reduced by the decreasing water availability in the roots as a result of induced water deficit stress. Miao *et al.* [52], provided evidence that verified nodule and *Rhizobium* sensitivity to water stress. Other related studies include those of Mangena [24], Ramos *et al.* [53], and Shetta [54] which demonstrated that anatomical changes or reduced nodulation during drought stress could cause a reduction in other various growth aspects such as stem height, stem xylem diameter, root dry matter and the entire plant biomass.

CONCLUSION AND FUTURE PROSPECTS

Drought stress has been widely reported to be a major abiotic stress factor, which drastically reduces the growth and yield of soybean plants. Low water levels in the soil cause a reduction in water availability for the roots and thus, decreasing water potential in the cells, which triggers a cellular hypertonic state. To continue their growth and development under drought stress, soybean plants need to adapt to these high solutes state through osmoregulation, as well as anatomical and morphological changes for long-term regulation [55]. The reduced percentage of total chlorophyll is another parameter that indicates water stress sensitivity in many plant species. This study found that the higher the concentration of chlorophyll, including other metabolites such as total phenolics and total flavonoids under water deficit, the better the cultivar performance. The improvement of these primary and secondary metabolite expression will enhance the antioxidant activity of water-stressed plants compared to cultivars, demonstrating poor antioxidant activity.

The size of the vascular tissue system and particularly, sclerenchyma, among the ground supporting the tissue system, will result in additional structural cell support during drought stress and altered functionality of the root nodules. Furthermore, this research demonstrated that highly reduced/ stunted stems emanating from limited water availability could be used as a morphological marker to identify low yielding soybean varieties. The inability of plants to reduce leaf surface area under water-limited environments can also be regarded as a morphological trait that contributes to low growth and yield capacity in soybeans. Meanwhile, low relative leaf water content (RLWC) also remains associated with low yielding soybean cultivars. Despite their great diversity in form, size and genetics, all legume crop plants carry out similar physiological processes in order to reproduce and cope with environmental stress. The tissues, organs and whole organisms show growth polarity derived from radial polarity of cell division of meristems, particularly serving to improve growth under different conditions [24, 55].

In the past few decades, a number of genetic engineering techniques were explored to propagate selected genotypes showing resistance to diseases and abiotic stress. The methods were usually coupled with *in vitro* propagation, through enhanced axillary shoot proliferation, node culture, *de novo* formation of adventitious shoots, zygotic and non-zygotic embryogenesis depending on the species and culture conditions [56]. Currently, the most frequently tested genetic manipulation tool, coupled with micropropagation methods for commercial production, utilises enhanced shoot proliferation from DNA bombarded cells or *Agrobacterium* infected meristematic cells [57]. As reported by Kane [58], the shoot culture method has consequently played an important role in the development of a industry worldwide that produces more than 350 million plants annually. Besides propagation, these culture systems, developed *in vitro* or *in vivo* have successfully produced pathogen-eradicated plants, assisted in the preservation of pathogen-eradicated germplasm and in the development of abiotic stress-tolerant plants, especially drought and salinity stress. If morphological and physiological markers related to plant growth and development are coupled with concepts such as genetic engineering to regenerate new high yielding plants, this can exert some level of control over effects of drought stress in the growth and development of soybean plants.

LIST OF ABBREVIATIONS

ABA   Abscisic acid

CM   Centimetres

DNA   Deoxyribonucleic acid

DPPH   2-Diphenyl-1-picrylhydrazyl

FANR   Food Agriculture and Natural Resources

LSA   Leaf surface area

RLWC Relative leaf water content

ROS   Reactive oxygen species

R3   Reproductive stage 3

V3   Vegetative stage 3

CONSENT FOR PUBLICATION

Not applicable.

CONFLICT OF INTEREST

The author declares no conflict of interest, financial or otherwise.

ACKNOWLEDGEMENTS

Declared none.

REFERENCES

[1]   Mattson JW, Haack RA. The role of drought in outbreak of plant eating insects. Bioscience 1987; 37(2): 110-8.
[http://dx.doi.org/10.2307/1310365]

[2]   Mangena P. The role of plant genotype, culture medium and *Agrobacterium* on soybean plantlets regeneration during genetic transformation. In: Khan SM, Malik A, Eds. Transgenic crops: Emerging trends and future perspective. London: Intech Open 2018; pp. 9-24.

[3]   Singh U, Singh B. Tropical grain legumes as important human foods. Econ Bot 1992; 46: 310-21.
[http://dx.doi.org/10.1007/BF02866630]

[4]   Kumar A, Pandey V, Shekh AM, Dixit SK, Kumar M. Evaluation of cropgro-soybean (*Glycine max I. [L] Merrill*) model under varying environmental condition. Amer-Eura J Agron 2008; 1: 34-40.

[5]   Liu F, Jensen CR, Anderson MN. Drought stress effect on carbohydrate concentration in soybean leaves and pods during early reproductive development: its implication in altering pod set. Field Crops Res 2003; 86: 1-3.
[http://dx.doi.org/10.1016/S0378-4290(03)00165-5]

[6]   Ishaq MN, Ehirim BO. Improving soybean productivity using biotechnology approach in Nigeria. World J Agric Sci 2014; 2(2): 13-8.

[7]   Liu S, Yan D, Wang J, Wang G, Yang M. Drought mitigation ability index and application based on balance between water supply and demand. Water 2015; 7: 1792-807.
[http://dx.doi.org/10.3390/w7051792]

[8]   Babovic J, Milic S. Irrigation effects in plant production in Serbia and Montenegro. In Program and Abstracts of the BALWOIS Conference on Water Observation and Information System for Decision Support, Ohrid, Republic of Macedonia.

[9]   Food Agriculture, Resources Natural. Issue No. 3 2017. Food, Agriculture and Natural Resources (FANR). Food security; early warning system. Directorate for the Southern African Development Community (SADC) 2017; (3):

[10]   Unger PW, Howell TA. Agricultural water conservation - a global perspective. J Crop Prod 1999; 2:

1-36.
[http://dx.doi.org/10.1300/J144v02n02_01]

[11]   Kovac L, Jakubova J, Sarikova D. Effect of tillage system and soil conditioner application on soybean (*Glycine max* (L.) Merrill) and its crop management economic indicators. Agri 2014; 60(2): 60-9.

[12]   Mangena P, Mokwala PW, Nikolova RV. Challenges of *in vitro* and *in vivo* Agrobacterium-mediated genetic transformation in soybean. In: Kasai M, Ed. Soybean- The Basis of Yield, Biomass and Productivity. London: Intech Open 2017; pp. 65-94.
[http://dx.doi.org/10.5772/66708]

[13]   Jauhar PP. Modern biotechnology as an integral supplement to conventional plant breeding: The prospect and challenges. Crop Sci 2006; 46: 1841-59.
[http://dx.doi.org/10.2135/cropsci2005.07-0223]

[14]   Yoshimura K, Masuda A, Kuwano M, Yokota A, Akashi K. Programmed proteome response for drought avoidance/tolerance in the root of a C(3) xerophyte (wild watermelon) under water deficits. Plant Cell Physiol 2008; 49(2): 226-41.
[http://dx.doi.org/10.1093/pcp/pcm180] [PMID: 18178965]

[15]   Cellier F, Conéjéro G, Breitler JC, Casse F. Molecular and physiological responses to water deficit in drought-tolerant and drought-sensitive lines of sunflower. Accumulation of dehydrin transcripts correlates with tolerance. Plant Physiol 1998; 116(1): 319-28.
[http://dx.doi.org/10.1104/pp.116.1.319] [PMID: 9499218]

[16]   Zidenga T. Progress in molecular approaches to drought tolerance in crop plants. Ohio: Ohio State University 2006.

[17]   Kisman M. Effects of drought stress on growth and yield of soybean. Indonesia: Bogor Agricultural University 2003.

[18]   Heschel MS, Riginos C. Mechanisms of selection for drought stress tolerance and avoidance in Impatiens capensis (Balsaminaceae). Am J Bot 2005; 92(1): 37-44.
[http://dx.doi.org/10.3732/ajb.92.1.37] [PMID: 21652382]

[19]   Izanloo A, Condon AG, Langridge P, Tester M, Schnurbusch T. Different mechanisms of adaptation to cyclic water stress in two South Australian bread wheat cultivars. J Exp Bot 2008; 59(12): 3327-46.
[http://dx.doi.org/10.1093/jxb/ern199] [PMID: 18703496]

[20]   Shinozaki K, Yamaguchi-Shinozaki K. Gene network involved in drought stress response and tolerance. J Exp Bot 2007; 58: 221–227.

[21]   Chen P, Sneller CH, Purcell LC, Sinclair TR, King CA, Ishibashi T. Registration of soybean germplasm lines R01-416F and R01-62581F for improved yield and nitrogen fixation under drought stress. J Plant Regist 2007; 2: 166-7.
[http://dx.doi.org/10.3198/jpr2007.01.0046crg]

[22]   Lobato AKS, Costa RCL, Neto CFO, *et al.* Morphological changes in soybean under progressive water stress. Int J Bot 2008; 4: 231-5.
[http://dx.doi.org/10.3923/ijb.2008.231.235]

[23]   Mangena P. Water stress: Morphological and anatomical changes in soybean (*Glycine max* L.) plants. In: Andjelkovic V, Ed. Plant, abiotic stress and responses to climate change. London: Intech Open 2018; pp. 17-40.
[http://dx.doi.org/10.5772/intechopen.72899]

[24]   Dekov I, Tsonev T, Yordanov I. Effect of water stress and high-temperature stress on the structure and activity of photosynthetic apparatus of *Zea mays* and *Heliathus annuus*. Photosynthetica 2000; 38: 361-6.
[http://dx.doi.org/10.1023/A:1010961218145]

[25]   Masuomi H, Darvish F, Daneshian J, Normohammadi G, Habibi D. Effects of water deficit stress on

seed yield and antioxidants content in soybean (*Glycine max* L.) cultivars. Afr J Agric Res 2011; 5: 1209-18.

[26]   Blokhina O, Virolainen E, Fagerstedt KV. Antioxidants, oxidative damage and oxygen deprivation stress: a review. Ann Bot 2003; 91(Spec No): 179-94.
[http://dx.doi.org/10.1093/aob/mcf118] [PMID: 12509339]

[27]   Pinheiro C, Chaves MM. Photosynthesis and drought: can we make metabolic connections from available data? J Exp Bot 2011; 62(3): 869-82.
[http://dx.doi.org/10.1093/jxb/erq340] [PMID: 21172816]

[28]   Taiz L, Zeiger E, Moller IM, Murphy M. Plant physiology and development. London: Sinauer Associates 2015; pp. 756-60.

[29]   Hassanzadeh M, Ebedi A, Panahyan-Ekivi M, *et al.* Evaluation of drought stress on relative water content and chlorophyll content of sesame (*Sesamum indicum* L.) genotypes at early flowering stage. Res J Environ Sci 2009; 3: 345-50.
[http://dx.doi.org/10.3923/rjes.2009.345.350]

[30]   Stajner D, Popovic BM, Taski K. Effects of Y- irradiation on antioxidant activity in soybean seeds. Cent Eur J Biol 2009; 4: 381-6.

[31]   Sakihama Y, Cohen MF, Grace SC, Yamasaki H. Plant phenolic antioxidant and prooxidant activities: phenolics-induced oxidative damage mediated by metals in plants. Toxicology 2002; 177(1): 67-80.
[http://dx.doi.org/10.1016/S0300-483X(02)00196-8] [PMID: 12126796]

[32]   Xu B, Chang SK. Antioxidant capacity of seed coat, dehulled bean, and whole black soybeans in relation to their distributions of total phenolics, phenolic acids, anthocyanins, and isoflavones. J Agric Food Chem 2008; 56(18): 8365-73.
[http://dx.doi.org/10.1021/jf801196d] [PMID: 18729453]

[33]   Oseni OA, Okoye VI. Studies of phytochemical and antioxidant properties of the fruits of watermelons (*Citrullus lanatus*). J Pharm Biomed Sci 2013; 27: 508-14.

[34]   Mogotlane AE, Mokwala PW, Mangena P. Comparative analysis of the chemical compositions of indigenous watermelon (*Citrullus lanatus*) seeds from two districts in Limpopo Province, South Africa. Afr J Biotechnol 2018; 17(32): 1001-6.
[http://dx.doi.org/10.5897/AJB2018.16552]

[35]   Torres AM, Mau-Lastovicka T, Rezaainyan R. Total phenolics and high-performance liquid chromatography of phenolics in avocado. J Agric Food Chem 1987; 35: 921-5.
[http://dx.doi.org/10.1021/jf00078a018]

[36]   Marinova D, Ribarova F, Atanassova M. Total phenolics and total flavonoids in Bulgarian fruits and vegetables. J Uni Chem Met 2005; 40: 255-60.

[37]   Peiretti PG, Karamac M, Janiak M, *et al.* Phenolic composition and antioxidant activities of soybean (*Glycine max* (L.) Merr.) plant during growth cycle. Agro (B Aires) 2019; 9(153): 1-15.
[http://dx.doi.org/10.3390/agronomy9030153]

[38]   Odhav B, Beckrum S, Akula U, Baijnath H. Preliminary assessment of nutritional value of traditional leafy vegetables in KwaZulu-Natal South Africa. J Food Compos Anal 2007; 20: 430-5.
[http://dx.doi.org/10.1016/j.jfca.2006.04.015]

[39]   Malenčić D, Popović M, Miladinović J. Phenolic content and antioxidant properties of soybean (*Glycine max* (L.) Merr.) seeds. Molecules 2007; 12(3): 576-81.
[http://dx.doi.org/10.3390/12030576] [PMID: 17851412]

[40]   Mittler R, Vanderauwera S, Gollery M, Van Breusegem F. Reactive oxygen gene network of plants. Trends Plant Sci 2004; 9(10): 490-8.
[http://dx.doi.org/10.1016/j.tplants.2004.08.009] [PMID: 15465684]

[41]   Yordanov I, Velikova V. And Tsonev, T. 2003. Plant responses to drought and stress tolerance. Bulg J

Plant Physiol 2003; 1: 187-206.

[42]   Alamillo JM, Díaz-Leal JL, Sánchez-Moran MV, Pineda M. Molecular analysis of ureide accumulation under drought stress in *Phaseolus vulgaris* L. Plant Cell Environ 2010; 33(11): 1828-37.
[http://dx.doi.org/10.1111/j.1365-3040.2010.02187.x] [PMID: 20545885]

[43]   Zrenner R, Stitt M, Sonnewald U, Boldt R. Pyrimidine and purine biosynthesis and degradation in plants. Annu Rev Plant Biol 2006; 57: 805-36.
[http://dx.doi.org/10.1146/annurev.arplant.57.032905.105421] [PMID: 16669783]

[44]   van Heerden PDR, Kiddle G, Pellny TK, *et al.* Regulation of respiration and the oxygen diffusion barrier in soybean protect symbiotic nitrogen fixation from chilling-induced inhibition and shoots from premature senescence. Plant Physiol 2008; 148(1): 316-27.
[http://dx.doi.org/10.1104/pp.108.123422] [PMID: 18667725]

[45]   Purcell LC, King CA, Ball RA. Soybean cultivar differences in ureides and the relationship to drought tolerant nitrogen fixation and manganese nutrition. Crop Sci 2000; 40: 1062-70.
[http://dx.doi.org/10.2135/cropsci2000.4041062x]

[46]   Rajan SS. Textbook of practical botany. New Delhi, India: Anmol 2003; Vol. 1: pp. 13-8.

[47]   Ladrera R, Marino D, Larrainzar E, González EM, Arrese-Igor C. Reduced carbon availability to bacteroids and elevated ureides in nodules, but not in shoots, are involved in the nitrogen fixation response to early drought in soybean. Plant Physiol 2007; 145(2): 539-46.
[http://dx.doi.org/10.1104/pp.107.102491] [PMID: 17720761]

[48]   Russin WA, Evert RF. Studies on the leaf of *Populus deltades* (*Salicaceae*): morphology and anatomy. Am J Bot 1984; 71: 1398-415.
[http://dx.doi.org/10.1002/j.1537-2197.1984.tb11997.x]

[49]   Prince SJ, Murphy M, Mutava RN, *et al.* Root xylem plasticity to improve water use and yield in water-stressed soybean. J Exp Bot 2017; 68(8): 2027-36.
[http://dx.doi.org/10.1093/jxb/erw472] [PMID: 28064176]

[50]   Den Herder G, Van Isterdael G, Beeckman T, De Smet I. The roots of a new green revolution. Trends Plant Sci 2010; 15(11): 600-7.
[http://dx.doi.org/10.1016/j.tplants.2010.08.009] [PMID: 20851036]

[51]   Miao S, Jin J, Shi H, Wang G. Effect f short-term drought and flooding on soybean nodulation and yield at key nodulation stages under pot culture. J Food Agric Environ 2012; 10(3): 819-24.

[52]   Ramos MLG, Pearsons R, Sprent JI, James EK. Effect of water stress on nitrogen fixation and nodule structure of common bean. Pesq Agro Bras 2003; 38(3): 339-47.
[http://dx.doi.org/10.1590/S0100-204X2003000300002]

[53]   Shetta ND. Influence of drought stress on growth and nodulation of Acacia origena (Hunde) inoculated with indigenous rhizobium isolated from Saudi Arabia. Am-Eurasian J Agric Environ Sci 2015; 15(5): 699-706.

[54]   Keyvan S. The effects of drought stress on yield, relative water content, proline, soluble carbohydrates and chlorophyll of bread wheat cultivars. J Anim Plant Sci 2010; 8: 1051-60.

[55]   Benjamin JG, Nielsen DC. Water deficit effects on root distribution of soybean, field pea and chickpea. Field Crops Res 2006; 97: 248-53.
[http://dx.doi.org/10.1016/j.fcr.2005.10.005]

[56]   Caponetti JD, Gray DJ, Trigiano RN. Plant development and biotechnology. New York: CRC Press 2005; pp. 9-14.

[57]   Mangena P. A simplified *in-planta* genetic transformation in soybean. Res J Biotechnol 2019; 14(9): 117-25.

[58]   Kane ME. Shoot culture procedures. In: Trigiano RN, Gray DJ, Eds. Plant development and biotechnology. New York: CRC Press 2005; pp. 145-57.

CHAPTER 4

Effect of Liquid Maxiflo (*Azospirillum* spp.) and Trykoside (*Trichoderma* spp.) on the Growth and Yield of *Pisum sativum* L. and Other Selected Vegetables

Nwagu Rodney Mashamba[*]

Agricultural Research Council, Agronomy and Technology Transfer Department, Private Bag X82075, Rustenburg 0300, Republic of South Africa

Abstract: The challenge for agriculture is to address the need for adequate food provision and a sustainable future for crop productivity. The solution for increasing food production can probably only be obtained through the expansion of arable land by increasing irrigation practices, or by increasing harvestable yields on available land through the improvement of agricultural technology. With regard to the latter approach, field experiments were conducted to determine the effect of Maxiflo and Trykoside; liquid formulations of *Azospirillum* and *Trichoderma* based products, respectively, on vegetative growth and yield of *Pisum sativum* L., *Brassica oleracea* var. *capitata*, *Lactuca sativa*, *Solanum tuberosum*, *Lycopersicon esculentum* and *Triticum aestivum*. A randomized complete block design with six treatments (Control, Maxiflo, Trykoside, Maxiflo + Trykoside, ComCat® and Kelpak®) was applied in all cases. Maxiflo and Trykoside were applied either separately or together. Two commercially available natural biostimulants, *ComCat*® and *Kelpak*®, also served as positive controls. Results showed that peas were most responsive to treatment with the bio-products in terms of the increase in yield obtained compared to other crops. Growth of cabbage and lettuce were not affected, but a significant increase in head mass was observed. Combined treatment of Maxiflo and Trykoside increased fruit yield in tomatoes as well as tuber sizes in potatoes. However, in both crops, the total yield was not significantly affected, especially compared to the yield in peas. In wheat, root growth was stimulated significantly by treatment with Trykoside, but no significant yield increases were observed compared to peas.

Keywords: Peas, Maxiflo, Trykoside, *ComCat*®, *Kelpak*®, Vegetative growth, Yield components.

---
[*] **Corresponding author Nwagu Rodney Mashamba**: Agricultural Research Council, Agronomy and Technology Transfer Department, Private Bag X82075, Rustenburg 0300, Republic of South Africa; Tel: 012 427 9999; Fax: 086 725 8482; E-mails: mashambar@arc.agric.za & rondey.nwagu8@gmail.com

Phetole Mangena (Ed.)

INTRODUCTION

More than 850 million people in the developing world, of which over 200 million are children, are chronically undernourished, while an estimated 1 to 3 billion people worldwide do not receive sufficient quantities of nutrients that are needed on a daily basis. Considering the estimated growth in the world population by 2050, it is clear that the challenge of providing nourishment to humans is significant. In 2002, the World Food Summit recommitted itself to halve the number of hungry people by the year 2015 [1]. How honourable this objective might have been, the finding of a solution is not evident as it implies considering a number of factors, including a) the economic status of individuals as determined by employment and minimum wages and b) ways and means to improve agricultural productivity. In both instances, a considerable amount of research is inevitable. Further, population growth is a relative uncertain factor that has to be considered.

It is predicted that population growth will occur in large measures, especially in African developing countries where poverty is rife. The challenge for agriculture as a sector is, therefore, to address the need for adequate food provision and a sustainable future for crop productivity. The problem for the future seems to be related to the fact that a solution for increased food production can probably only be obtained in three possible ways, namely (i) through the expansion of arable land, (ii) increasing sustainable irrigation practices or (ii) by increasing harvestable yields through the improved agricultural technology. However, according to de Vries [2], severe soil erosion, especially in Africa, is minimizing the number of acreages available for cultivation, leaving an almost impossible task of increasing the amount of arable land. Further, most of the irrigatable soil on the planet is probably already utilized, and chances for expansion seem slim. This leaves the increase of crop yields on currently available land as the only and most likely alternative [3].

To obtain the latter goal of increasing crop yield, future agricultural research will have to focus on certain key areas. These include improved disease and pest control either through conventional or modern breeding for resistance against specific diseases or by improving chemical control methodology and technology, *e.g.*, by finding new effective but cheaper products for application by farmers in the developing world, and by applying natural bio-stimulants from plants either as seed treatments or as a foliar spray or both [4, 5]. The development of natural products to achieve this goal has gained support in the recent past. Previous studies showed significant increases in wheat yield when grown in mixed stands with corn cockle. A biostimulatory substance isolated from the corncockle, agrostemin, increased grain yields when applied to both fertilized and unfertilized land areas used to grow wheat [6]. Chopped alfalfa also had a stimulatory effect on the growth of a number of vegetables, and the active substance was later identified as triacontanol [7].

Saponins isolated from crude mung bean extracts were found to increase germination and also enhance the vegetative growth of cultivated mung beans [8]. The effective application of this knowledge can be instrumental in increasing crop yields and contributing towards food security, especially in developing countries. Underlying the need to develop new cheaper natural products is the fact that the lack of an efficient integrated disease-weed-pest management system has been identified as one of the main reasons for inadequate food production in Africa and other developing countries. Further, in developed countries, increased resistance by consumers to purchase plant products grown from either genetically manipulated crops or crops treated with synthetic chemicals is currently experienced [9]. Legislation restricting the use of many synthetic crop protectants in recent years as well as the banning of copper-containing synthetic pesticides in Europe, has led to increased organic farming practices [10].

This means that indispensable tools used in crop production systems may be eliminated without existing alternatives. This prompted research activities towards developing natural products as alternative crop protectants in recent years and accelerated the search for natural chemicals from plants, also known as green chemicals [9]. A recently established company, Axiom Bio-Products Pty Ltd, manufactured two products, namely Maxiflo *(Azospirillum* based) and Trykoside *(Trichoderma* based) in liquid form. The rationale for this study was to investigate the possibility of increasing growth and yield in six economically important crops by treatment with Maxiflo and Trykoside, as representative of bio-stimulatory agents in comparison with two commercially available natural bio-stimulants, *ComCat*® and *Ke/pal<*®, serving as positive controls. The objectives were to determine the effect of Maxiflo and Trykoside on the growth and yield of *Pisum sativum* compared with other selected crops both separately and in combination.

CULTIVATION AND ECONOMIC IMPORTANCE OF *PISUM SATIVUM*

Peas *(Pisum sativum)* is a cool-season annual crop adapted to semi-arid climates and belongs to the Fabaceae family. Peas have the capacity to fix atmospheric nitrogen into the soil so that it can be available for utilization by other plants (Fig. **1**). Peas are generally considered as a low fertility crop that thrives well on fertile soil. This crop requires a cool, relatively humid climate, and it is grown at higher altitudes in the tropics with temperatures ranging from 7 to 30°C [11]. The optimum temperature levels for vegetative and reproductive periods of peas were reported to be 21 and 16°C, and 16 and 10°C (day and night), respectively [12].

The optimal planting dates for peas range from mid-April when soil temperatures are above 4.44°C to mid-May. Peas are very sensitive to drought and grow best in regions of moderate rainfall or with irrigation.

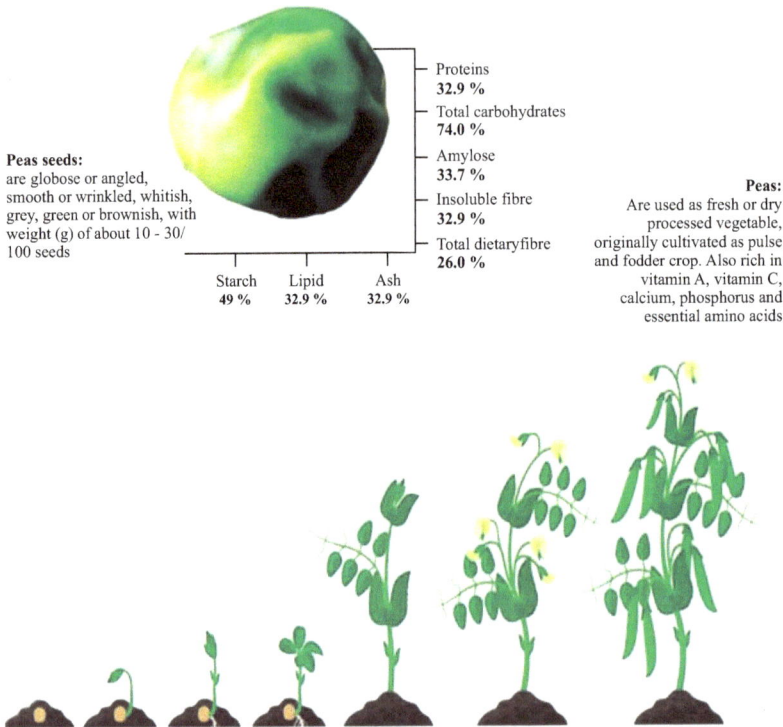

Peas seeds:
are globose or angled, smooth or wrinkled, whitish, grey, green or brownish, with weight (g) of about 10 - 30/ 100 seeds

Proteins
32.9 %
Total carbohydrates
74.0 %
Amylose
33.7 %
Insoluble fibre
32.9 %
Total dietary fibre
26.0 %

Starch   Lipid   Ash
49 %   **32.9 %**   **32.9 %**

Peas:
Are used as fresh or dry processed vegetable, originally cultivated as pulse and fodder crop. Also rich in vitamin A, vitamin C, calcium, phosphorus and essential amino acids

Fig. (1). Proximate composition of protein, carbohydrates, lipid, starch and fibre of pea seeds.

Peas can also be grown successfully during mid-Summer and early fall in those areas with relatively low temperatures and good rainfall or where irrigation is practiced. For very early crops, a sandy loam is preferred; for large yields where earliness is not a factor, a well-drained clay loam, sandy loam or silt loam is preferred [13]. It also requires a pH of 6.5 or higher for maximum yield. Peas can also be grown in a no-till or conventional tillage cropping system, and it requires a high amount of moisture for germination than cereal grains. Pea growing season varies from 80-100 days in semi-arid regions, and it can reach up to 150 days in humid and temperate areas [11]. The association of pea with nitrogen-fixing bacteria contributes to the development of low-input farming systems. Total pea production costs incurred have been lower in small and medium-scale farmers due to lower use of inputs by these farmers [14].

This crop is among the legumes that contribute about 27% of the world's economic crop production after cereals. As a critical source of proteins, fibre and essential constituent of sulphur -containing amino acids such as cysteine and methionine, pea play a vital role in providing food and fodder for humans and animals (Fig. **2F**). Pea production continued to increase from 1.3 million tonnes to 3.2 million tonnes in the previous years [14]. Pea cultivation requires high economic inputs, for purchasing of larger dosages of manure and fertilisers, labour, and pesticides. However, this crop has the potential to generate billions of dollars for farmers motivated to diversify to more remunerative cropping systems than using traditional, less profitable cultivation patterns.

OTHER SELECTED VEGETABLE CROPS

Cabbage (*Brassica oleracea*)

Cabbage is a member of the Brassicaceae family, the same family as broccoli, Brussel sprouts, cauliflower, kale, green mustard, and collards (Fig. **2A**). Collectively, these crops are referred to as mustards. Cabbage is well adapted for growth in cool climates. This is a popular vegetable worldwide because of its adaptability to a wide range of climate and soil, its ease of production and storage, and its food value [15]. Cabbage is a cool-season biennial crop that is grown as an annual vegetable requiring 60 to 100 days from sowing until market maturity depending on the cultivar. The ideal monthly temperature for optimal growth of cabbage ranges from 15 to 18°C. Temperature greater than 24°C induces bolting in cabbage, but cultivars differ in their susceptibility. Cabbage has been used as a food crop since antiquity [16]. Cabbage does well in a relatively cool, moist climate. For this reason, cabbage is cultivated in the north parts of South Africa, mainly in autumn, winter, and spring. The optimum temperature for growth and development on average is approximately 18°C, with an average maximum of 24°C and an average minimum of 4.5°C [17].

It is also fairly resistant to frost and readily survives minimum temperatures as low as -3°C without noticeable damage. Optimum temperature and humidity are seldom encountered, but cabbage, fortunately, has wide adaptability. It can consequently be cultivated in most areas throughout the year, although quality and yield are usually poor during the summer months because of the high infestation of pests. Cole crops do best in well-drained, fertile loam, but they can be successfully grown on a wide range of soils, provided that drainage and fertility are good. For summer and early winter planting, cabbage does best on the heavier loams, while the spring crop does best on a sandier loam. As these vegetables respond readily to organic fertilisers, it is recommended that adequate organic material be incorporated in the soil [18]. Cole crops do best in soil with a pH value between 6.0 and 6.5. If the value is below 5.8, it is advisable to apply lime and the type and amount of lime will be indicated by soil analysis.

Lettuce (*Lactuca sativa*)

Lettuce *(Lactuca sativa)* belongs to the *Asteraceae* (sunflower or daisy) family (Fig. **2B**). It has a small cylindrical root system, with an effective root width of 25 cm, which implies that the plants should be closely spaced (both in and between rows). It is an annual plant native to the Mediterranean area cultivated as early as 4500 BC, initially for the edible oil extracted from its seeds. Salad lettuce became popular with the ancient Greeks and Romans [19]. Lettuce is a cool-season crop and grows best within a temperature range of 12–20°C. It is an annual plant closely related to the common wild or prickly lettuce weed [20]. It is so sensitive to low temperature provided there is high elevations during summer. Lettuce grows well on a wide variety of soils, provided climatic requirements are met.

Cultivated lettuce was derived from the wild or prickly lettuce *Lactuca sioriola*. There are five types of lettuce, namely crisp head, butter head, romaine, loose leaf or bunching and stem lettuce [19]. Lettuce is currently an economically important crop grown in large quantities all over the world. The leaf colour of commercial lettuce cultivars varies from yellow-green to dark red. Head lettuce grows best at temperatures between 15 to 10°C. Warm sandy soils are preferred for the early harvestable types, while loam to clay loam or peat are suited for lettuce produced later in the season [19]. The part that contains the highest nutritional value is the dark green outer leaves despite the fact that in calories, it is low. Each head contains only 65 to 70 kilocalories [21]. The use of lettuce includes extraction of oil from its seeds and use in salad and relish.

Wheat (*Triticum aestivum*)

All wheats, whether wild or cultivated, belong to the genus *Triticum* (Fig. **2D**). Wheat is a cereal grain crop classified under the *Poaceae* or grass family. Its complete botanical classification is as follows: genus: *Triticum,* species: *aestivum* and *turgidium.* The species are categorized into groups, *aestivum* for bread wheat; *compactium* for common wheat; *spelta* for spelt wheat and *turgidium, poulard* (branched) wheat and *durum* for hard wheat. Wheat is the most important world crop today, judging from the land area under production [22]. Wheat has a relatively broad adaptation, being very well adapted to harsh climates, and will grow well where rice and maize cannot. Generally, the winter climate of a particular area determines whether winter or spring types are grown. Wheat is grown on a wide range of soils in temperate climates where annual rainfall ranges between 30 and 90 cm. Such areas constitute most of the grasslands of the world's temperate regions. Many of these soils are deep, well-drained, dark coloured, fertile, and high in organic matter, and they represent some of the world's best soils. Loam to sand loamy soils are ideal for planting of wheat crops [23].

Potato (*Solanum tuberosum*)

Potato is a herbaceous plant belonging to the Solanaceae family (Fig. **2E**). Other well-known crops belonging to the same family are tomato (*Lycopersicon esculentum*), the eggplant (S. *melongena),* various species of chilipeppers *(Capsicum)* and tobacco *(Nicotiana tabacum).* Potato is also classified under the hemispherical type of root system. Potatoes are normally cultivated under temperate climates or the mountains of tropical areas. Amongst all tuber crops, potato top the list in terms of hectares under cultivation, followed by cassava and sweet potatoes. Potato tubers give an exceptionally high yield per hectare, many times that of any grain crop and are used as processed food and livestock feed [24, 25]. The potato may be classified as a dicotyledonous annual, although it can persist in the field as tubers from one season to the next. Potato is a cool-season crop, slightly tolerant of frost, but easily damaged by freezing weather near maturity. Today potato encircles the globe, they are grown on every continent and can also be planted as soon as the soil temperature reaches about 5°C, while the emergence is more rapid at 20 to 22°C.

Soil temperatures of 15 to 18°C appear to be the most favourable for common potato varieties. For example, in varieties of the *Tuberosum* subspecies, short days and moderate temperature, particularly low night temperatures, stimulate tuber initiation, however, tubers mature late under short days [25]. Maximum yields of high-quality tubers are produced when the mean temperature is between 15°C and 18°C during the growing season. Tuberisation (tuber formation) is also favoured by long days of high light intensity. Optimum temperature for tuber development is about 18°C. Tuberisation is progressively reduced when night temperatures rise above 20°C and are totally inhibited at 30°C. The potato crop develops best on deep, friable soil that have good water retention, because it has a relatively weak root system, impermeable layers in the soil limit rooting depth, which in turn restricts the availability of water to the plant during dry period [26].

Tomato (*Lycopersicon esculentum*)

Tomato also belongs to the plant family Solanaceae, is a native of tropical America, and is also classified as an annual season under the plant group with a hemispherical type of root system (Fig. **2C**). Tomato is a warm-season plant that requires three to four months of sunshine from the time of seeding up to the production of the first ripe fruit [27]. It thrives best when the weather is clear and rather dry, and temperatures are uniformly moderate (18 to 29°C). Tomato can be cultivated on nearly all types of soils, although light, well-drained and fertile soil is best suited for producing early fruit of high quality. Loams and clay loams have a greater water holding capacity and are well suited for producing tomatoes at a pH ranging from 5.5 to 7.0. World production of tomato has increased to approximately 10% since 1985, reflecting a substantial increase in dietary use of the crop. Nutritionally, tomato is a significant dietary source of vitamin A and C. Further, recent studies have shown the importance of lycopene, a major component of red tomatoes with strong antioxidant properties, which reduces the incidence of several cancer types [27].

Fig. (2). Examples of the production fields showing the selected vegetable crops used for examining the effect of bio-stimulants of *Azospirillum* spp. and *Trichoderma* spp. culture used for enhanced growth and yields. **(A)** *Brassica oleracea* var. *capitata*, **(B)** *Lactuca sativa*, **(C)** *Lycopersicon esculentum*, **(D)** *Triticum aestivum*, **(E)** *Solanum tuberosum* and **(F)** *Pisum sativum* L.

USE OF PLANT EXTRACTS AS BIO-STIMULANTS

The isolation and purification of active compounds from plants have proved beneficial and currently received considerable attention in research. However, this recognition may place them in the same category as synthetic chemicals in terms of production costs and even their impact on the environment. Hence, the application of entirely plant-based chemicals may be a feasible alternative due to the fact that they are bio-degradable and environmentally friendly compared to the traditional synthetic agrochemicals [9]. However, the effective application of plant-based bio-stimulants during agricultural practices has only been established in a few cases emphasizing the necessity for additional research. Soil contains many microbes, including beneficial ones that are essential to good crop growth. Recent research has begun to show how to manage soil microflora to favour plant growth and yields.

One approach has been to add some of the best ones to the fields in order to create a more favourable soil microbe environment. However, most of these introductions failed because the native microflora is more competitive than the introduced ones. Microbials are safe alternatives to the use of chemical pesticides [11, 12]. The application of microorganisms such as fungal *Trichoderma* and bacterial *Azospirillum* spp. to soil have the potential to make a meaningful contribution in achieving high productivity and bio-control. The inoculation of plants with *Azospirillum* has shown significant changes in various plant growth parameters that may affect crop yields [28, 29]. Based on worldwide field data accumulated over the past 30 years, a strong indication exists that *Azospirillum* is capable of promoting the yield of important crops in different soils and climatic regions.

These data showed significant yield increases in some published reports. *Azospirillum* has shown a positive influence on plant growth, crop yield and nitrogen content of the plant under certain environmental and soil conditions [30]. Review data from field inoculation experiments with *Azospirillum* spp. showed significant increases (5 to 30%) in the yield of published reports. The benefits observed from *Azospirillum* inoculation were mainly improved root development and enhanced water and mineral uptake. Available evidence indicates that secretion of plant-growth-promoting substances by the bacteria is at least partly responsible for these effects. During recent years, researchers have been focusing on the production of plant growth, promoting substances by this bacterium (*Azospirillum*) as a possible mechanism for the observed plant growth promotion [29].

Trichoderma Species and their Potential Application in Agriculture

A *Trichoderma* fungus species forms the basis of the product Trykoside. Five species have been described namely, *T. harzianum; T. koningii; T. longibranchiatum; T. pseudokoningii* and *T. viride* that are widely distributed in the soil, plant material, decaying vegetation, and wood (Fig. **3**). *Trichoderma* species are generally found as dominant components of the microflora in most soil types, including the forest humus, agricultural and orchard soils [31]. It is reported to grow on living plants and is not associated with plant diseases. There is a great diversity between the genotype and phenotype of wild strains, but they are all highly adapted and may be heterocaryotic (*i.e.*, contain nuclei of dissimilar genotype within a single organism and, hence, are highly variable). *Trichoderma* spp. are used in food and textile industries, and are highly efficient producers of many extracellular enzymes. They are used for the production of cellulases and other enzymes that degrade complex polysaccharides [30].

The enzymes are used in poultry feed to increase the digestibility of hemicelluloses from barley or other crops. Interestingly, *Trichoderma* spp. are also used as bio-control agents, and with or without legal registration for the control of plant diseases and plant growth promoters in the agricultural industry. From an agricultural perspective, the biological control of soil-borne plant pathogens with *Trichoderma* spp. has been well documented [32–35]. *T. harzianum* has been extensively used as a bio-control agent due to its capability to control a large variety of phytopathogenic fungi that are responsible for major crop diseases [36]. *Trichoderma* species have provided varying levels of biological control of a number of important soil-borne pathogens, including *Phytophthora cactum*, *Pythium* spp. and *Verticillium dahliae*. Isolates of *T. harzianum* have been reported as antagonists of mycelia or sclerotia of these soil-borne pathogens. *T. harzianum* formulated in alginate pellets colonized sclerotia of *Sclerotium sclerotiorum* under laboratory and field conditions [37 - 40]. Specific strains in the genus *Trichoderma* colonise and penetrate plant root tissues and initiate a series of morphological and biochemical changes in the plant, considered to be part of the plant defense response, which in the end leads to induced systemic resistance (ISR) of the entire plant [41].

According to Sharon *et al.* [44], *T. harzianum* was effective in the control of other diseases, some of which are caused by the pathogen *Rhizoctonia solani, Sclerotinia minor, Fusarium oxysporum, Sclerotium rolfsii* and some *Pythium* and *Phytophthora species.* Alone or in combination with other *Trichoderma* spp., it is regarded as the active component of several products inhibiting the growth of fungal plant pathogens. The involvement of *Trichoderma* species in ISR of crops such as cotton has also been reported [45]. Induced systemic resistance implies that the treatment of crops enhances the resistance of crops towards fungal infection by activating the natural mechanisms within the plants. Protease production by *T. harzianum* has also been associated with bio-control of the root-not nematode *Meloidogyne javanica* on tomato plants. Importantly, *Trichoderma* species are often able to suppress the growth of endogenous fungi on agar medium and therefore mask their presence. As a result, the routine use of bio-control agents for controlling plant diseases in agriculture has not been realized.

There is one feature that could make such agents more attractive and that is the possibility of enhanced crop growth in addition to disease control. Baker *et al.* [46] reported that such enhancement had been achieved with *T. harzianum*. Recently Jing *et al.* [47] showed that *Trichoderma* application had alleviated pathogen infection while promoting plant growth, the root-colonizing ability, yield and quality of lettuce. This information emphasises the dual use of these species as bio-control and bio-stimulant agents in the event that commercial synthetic fungicides are banned or unable to solve the problem. Finally, Harman [48] observed that *Trichoderma* spp. are favoured by the presence of a high level of plant roots, which they colonize. Some *Trichoderma* strains are highly competent in the rhizosphere, where they colonize and grow on roots while contributing to root development. Thus, if applied as a seed treatment, the best strain will colonize root surfaces and be a useful application in enhancing plant and root growth [49].

Azospirillum Species and their Potential Application in Agriculture

Azospirillum bacterium, on the other hand, forms the basis of the product Maxiflo and belongs to the Azotobacteraceae family (Fig. **3F**). It is an aerobic bacterium meaning that it requires oxygen to play its role in the soil. These microorganisms are characterized by their high nitrogen-fixing ability (diazotrophs) and are found in abundant numbers in the rhizosphere as well as in the intracellular spaces of the roots of certain cereals and legume crop plants [50]. It lives in close proximity to plant roots (*i.e.*, in the rhizosphere or within plants). As is the case for Trykoside, Maxiflo is also manufactured in liquid form. *Azospirillum* living in association with roots of cereal grain has been reported to stimulate growth. This relationship is viewed as associative symbiosis in which bacteria receive non-specific photosynthate carbon from the plant and, in turn, provide the plant with fixed N_2, hormones, signal molecules, vitamins, iron, *etc* [50, 51].

Gaskins & Hubbel [52] confirmed an increased growth rate in *Pennisetum americanum* after inoculation with *A. brasilense*, strain 3t, as compared to treatment with kinetin and GA_3 used as positive controls. The inoculation of pearl millet with *A. brasilense*, strain S14, has resulted in significant increases in growth and dry matter production under both sterile and non-sterile conditions. The beneficial effects of *Azospirillum* species on plant development are attributed to the production of phytohormones [50, 53]. Inoculation of rice with *Azospirillum* has even been suggested as an alternative to chemical fertilization. The plant stimulatory effect exerted by *Azospirillum* has been attributed to several mechanisms, including biological nitrogen fixation (BNF) and production of plant growth-promoting substances, also known as PGRs (plant growth regulators) [54 - 56]. Environmental factors such as O_2 partial pressure and mineral nitrogen concentration have been reported to influence the process of N_2 fixation in *Azospirillum* [57].

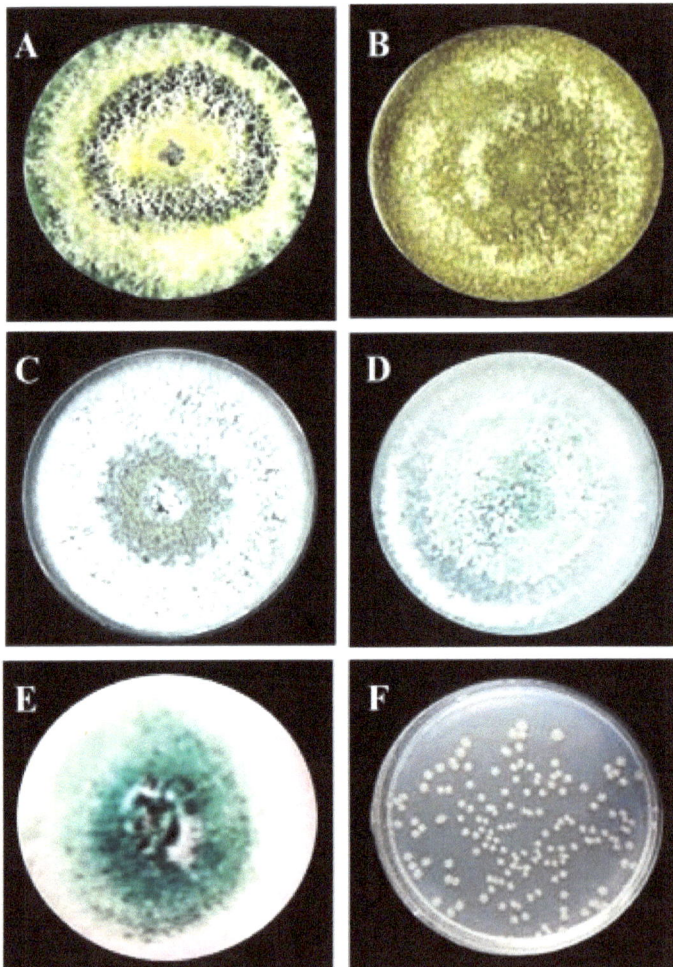

Fig. (3). Examples of variations and morphology of some of the *Trichoderma* and *Azospirillum* species shown on agar plates. **(A)** *T. harzianum*, **(B)** *T. koningii*, **(C)** *T. longibranchiatum*, **(D)** *T. viride*, **(E)** *T. pseudokoningii*, and **(F)** *A. brasilense* [42, 43].

Agricultural applications of *Azospirillum* spp. are commonly limited by low concentrations of assimilative carbon in the field. However, the use of *Azospirillum* spp. and other free-living nitrogen-fixing bacteria represents an enormous opportunity for agriculture as plant-growth-promoting rhizobacteria [50]. The beneficial impact of bacterial N_2-fixation on plant growth appears to be less significant than that of the rhizobia-legume symbiosis [29, 50, 58]. However, N_2-fixation remains important for bacterial survival in N-poor soils and possibly in the root environment. Improved nitrogen fixation resulted in an increased bacterial population on roots and consequently increased plant growth. Furthermore, inoculation of crop plants or the seeds of crop plants with *Azospirillum* increased the number of lateral roots and root hairs, thus enhancing the uptake of nutrients through increased root surface [59].

Azospirillum brasilense increased the growth and yield of tomato plants. Recently Kenny [60] confirmed that an *Azospirillum* spp. significantly reduced the occurrence of diseases in tomatoes and green peppers and simultaneously increased both plant size and yield under field conditions. Most studies of the *Azospirillum* plant association have been conducted on cereals and other grasses. However, a few field studies on oats, sorghum and other crops under appropriate growth conditions confirmed increases in plant dry mass and yield due to *Azospirillum* inoculation. From an agricultural perspective, the confirmation of growth improvement and yield increases in crop plants due to *Azospirillium* inoculation have been well documented. The inoculation of wheat with *A. brasilense* increased the rhizosphere population of *Azospirillum* and increased plant height, dry weight, nitrogenase activity and grain and straw yields.

In a field experiment in India, nitrogen fertilizer applications (up to 120 kg N/ha) and inoculation of seed with *A. brasilense* and *Azotobacter chroococcum* showed a significant increase in dry matter production, grain yield and grain protein content of wheat. Similar results have been published on sorghum and maize [61 - 63]. The optimization of the *in vitro* production of potatoes in South Africa using cytokinin and other PGRs has also been reported to have strong promotive effects on tuberization and constitutes the tuberization stimulus, either alone or in combination with other substances [64]. However, according to Leclerc *et al.* [64], PGRs failed to induce tuberization when sucrose supply was inadequate. It has been stated that the use of growth retardants rather than bio-stimulants has improved the microtuber formation of potato. According to Forti, *et al.* [65] tuberization in potatoes depends on the genotype. More recently, reports that *Azospirillum* has the ability to protect plants against various stresses, *e.g.*, drought stress through adjusting the turgor in cells, have become available. This aspect needs more consideration from an agricultural perspective [65, 66].

CURRENT TRENDS IN CROP GROWTH AND YIELD IMPROVEMENT

Before dealing with the role antagonistic soil micro-organisms can play in crop production systems, it is necessary to take note of existing technology to improve the growth and yield of agricultural and horticultural crops. Firstly, seeding rates and planting dates have always been the simplest measure to manipulate crop yields. However, a thorough knowledge of crop cultivars in terms of optimal planting dates as well as optimal environmental conditions necessary for optimal production, is essential. Secondly, the use of fertilizers as a measure to manipulate crops has been practised for centuries in most countries where agriculture is well developed, but the essential role of fertilisers in modern farming has become clear only during the last 50 years. In 1939 the world's farmers used 9 million tons of plant nutrients (mainly N, P and K), while in 1970, about seven times as much was used [67]. The application of plant nutrients was essential to support the agricultural revolution, which began in many temperate countries and had a great influence on the production of crops.

At present, bio-fertilization accounts for approximately 65% of the nitrogen supply to crops worldwide. Legumes were often used as green fertilizers in the past due to their nitrogen-fixing ability. The bacterial strains that are most efficient in this regard belong to the genera *Rhizobium, Sinorhizobium, Mesorhizobium, Bradyrhizobium, Azorhizobium* and *Allorhizobium* and are those strains that have been studied in most detail [27]. One of the recent approaches to bio-fertilisation is to apply natural bio-stimulants such as *Seagro*®, *Kelpak*® and *ComCat*® together with normal fertilizers as a means to enhance plant growth and productivity on existing arable land. These products are normally applied as foliar sprays, but they can also be applied as seed treatments [14]. Similar objectives with foliar sprays of soil micro-organisms on agricultural crops that include *Trichoderma* and *Azospirillium,* have been set in recent research projects with the aim to test their application potential in agriculture. This approach prompted this study.

EFFECT OF LIQUID MAXIFLO AND TRYKOSIDE ON PEA GROWTH AND YIELD

Pea (cultivar Solara) seeds were planted in moist soil. The planting density for peas was estimated at 35.556 plants per hectare with 4 cm inter-row spacing and 75 cm intra-row spacing in plots, with a plot size of about 5.625 m². *ComCat*®, a commercial bio-stimulant of plant origin, was used as a positive control in these two field trials (Table 1). The manufacturers claim that the product promotes plant growth and development, physiologically assists, and strengthens the plants' natural resistance mechanisms to pathogen attacks and encourages root growth

(Agraforum, Germany). The effects of Trykoside and Maxiflo were tested both separately and in combination with the growth and yield of peas and other vegetables, as indicated in Table 1 with minor variations per crop.

Suspensions of Maxiflo and Trykoside, as well as *ComCat®*, were applied at the rate of 890 L ha^{-1} for peas. Pods were harvested four times by hand, 13, 14, 15 and 16 weeks after planting during the 2003 growing season. The number of pea pods per plot was counted 13, 14, 15 and 16 weeks after planting, respectively. This was added to calculate the total number of pods per hectare. The fresh mass of pods containing seeds and seed fresh mass separately were measured per plot at harvest on the 13, 14, 15 and 16 weeks after planting. Analysis of variance was done on all parameters used to quantify the vegetative growth and yield components to determine the significance of differences between means using the NCSS 2000 program. Tukey's least significant difference (LSD) procedure was employed to separate means at the 5% ($P<0.05$) level.

Table 1. Summary of the treatments applied to the plant fields.

Treatments	Amount	Spray Information
Maxiflo	1 L/hectare	Spray at planting and three sprays every 2 weeks thereafter (peas),
Trykoside	1 L/hectare	Spray at planting and three sprays every 2 weeks thereafter (peas)
Maxiflo-trykoside	1 L each/hectare	Spray at planting and three sprays every 2 weeks thereafter (peas)
Comcat◦	100 g/hectare (at 10:90 ratio)	Spray at planting and three sprays and every 4 weeks thereafter (peas),
Comcat◦ -maxiflo-trykoside	100 g ComCat + 1 L each/hectare of Maxiflo-Trykoside	Spray at planting and three sprays every 4 weeks thereafter (peas),
Control	None	None

Yield Components

Maxiflo and Trykoside applied separately as well as together had a significant reducing effect on the pod number compared to the untreated control (Fig. 4). On the other hand, although *ComCat®* applied separately had only a slight enhancing effect on pod number, it contributed to a significant increase (excess of 20,000 pods/ha) in pod number compared with when applied in combination with Maxiflo and Trykoside. Interestingly, although Maxiflo applied separately had a reducing effect on total number of pods (Fig. **4A**) it contributed to a significant increase in both pod fresh weight (Fig. **4B**) and seed fresh weight (Fig. **4C**).

Trykoside applied separately showed the same tendency to reduce both pod and seed fresh weight, as was the case for pod number. Maxiflo and Trykoside, in combination, as well as ComCat®, applied separately, showed the same tendency towards reducing pod fresh weight but increased the seed fresh weight.

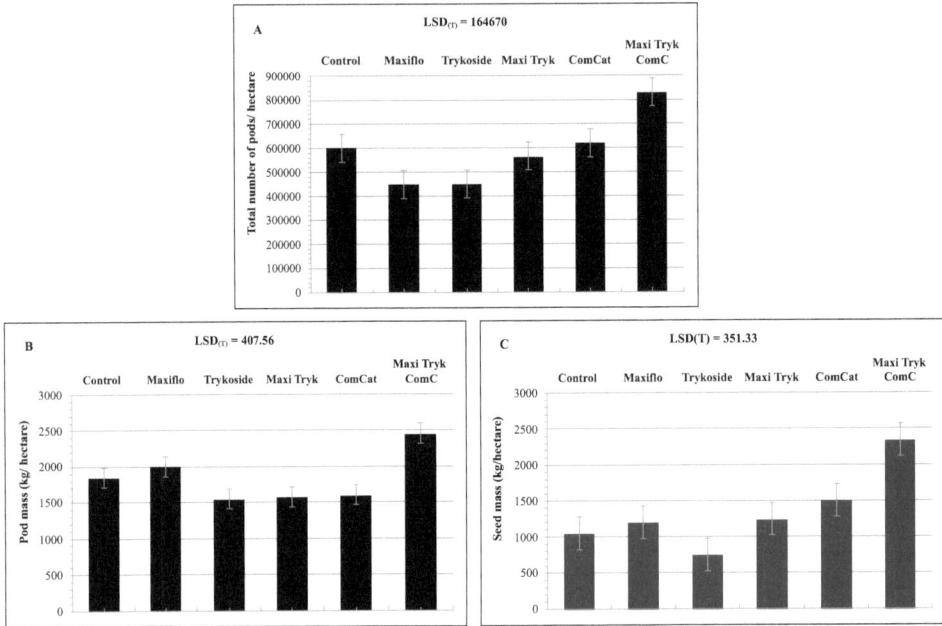

Fig. (4). The effect of Maxiflo and Trykoside both separately and in combination with yield components (number of pods, pod mass and seed mass/ hectare) of *Pisum sativum* L.

This was significant in the case of *Comcat*® compared to the untreated control. However, when Maxiflo, Trykoside and *ComCat*® were applied in combination, a sharp and statistically significant increase in seed fresh mass as it was observed for pod number prevailed. In peas, the three-way Maxiflo + Trykoside + *ComCat*® combination treatment significantly increased pod number as well as pod and seed fresh mass while *ComCat*® applied separately and Maxiflo + Trykoside applied as a two-way combination treatment only resulted in increased seed mass. It was expected that an increase in pod number would result in an increase in seed number and mass. Results implied that the three-way combination treatment must have had an enhancing effect on flower formation and subsequently pod number to achieve this increase. Interestingly, *ComCat*® alone as well as the Maxiflo + Trykoside two-way combination treatment did not have the same effect on pod number, making it difficult to ascertain which of these products had an effect on flower formation and pod yield increases.

It has been reported that the brassinosteroids (BRs) contained in *ComCat*® act as the active substance but also that BRs stimulate flower formation in a number of crops [6]. However, as *ComCat*® applied separately did not have a significant effect on pod numbers in peas. It seemed that the effect on flower formation reported on other crops is not applicable to a legume crop such as pea. This might indicate that, depending on the foliar application of these products either separately or in combination, a different metabolic-related mechanism of action is triggered in the plants. Furthermore, a combination of treatments might indicate a synergistic effect between products in the case where the combination treatments showed a tendency to enhance the pea yield or an antagonistic effect in the case where combination treatments resulted in the opposite. By means of a metabolic approach, this aspect needs further investigation.

On the other hand, compared to the untreated control, Maxiflo and Trykoside supplied separately rather inhibited the total number of pods counted over four harvests while the addition of *ComCat*® in a three-way combination increased the pod number. From this, it seems reasonable to postulate that the commercial bio-stimulant might have prevented the abscission of pods, a phenomenon that is well known in legume crops such as pea and beans, leading to the increased pod number observed in this study [68]. This is in agreement with the report of Molahlehi [69] on the reducing effect of *ComCat*® on pod abscission in dry beans under glasshouse conditions. Interestingly, notwithstanding the effect of the products under scrutiny on pod number, Maxiflo and *ComCat*® supplied separately as well as the Maxiflo + Trykoside two way and Maxiflo + Trykoside + *ComCat*® three-way combination treatments had a significant enhancing effect on the seed fresh mass in peas. It can be speculated that the increase in seed mass was achieved *via* enhanced carbohydrate photosynthate translocation from the storage organ (source) to the developing seed (sink) during the grain filling stage leading to increased protein levels *via* conversion from carbohydrates. It has been reported that brassinosteroids, the active compound of *ComCat*®, has a membrane energizing effect enhancing sucrose translocation over membranes as well as the source: sink relationship [70]. However, for Maxiflo and Trykoside, this mechanism of action is unknown.

A few reports on the effect of the *Azospirillum* (bacterium) based product, Maxiflo, on the yield of legumes are available in the literature and none on the *Trichoderma* (fungus) based product Trykoside. Field inoculation of garden peas and chickpeas with *A. brasilense* produced a significant increase in seed yield but did not affect the dry matter yield [71]. The author suggested that the increase in yield was the result of increased nitrogen fixation after inoculation with *Azospirillum*. Besides its nitrogen-fixing ability, *Azospirillum* spp. secretes phytohormones such as auxins, cytokinins and gibberellins [63]. Of these, auxin is quantitatively the most abundant phytohormone secreted by *Azospirillum,* indicating that auxin production, rather than nitrogen fixation, might be the major factor responsible for the stimulation of rooting and, hence, enhanced plant growth and yield. The inoculation of legumes with *Azospirillum* alone in the case of naturally-nodulated legumes, were shown to benefit plant growth under both greenhouse and field conditions [68].

GROWTH ANALYSIS OF OTHER GRAIN AND VEGETABLE CROPS

Cabbage

Statistically significant differences in plant diameter were observed only during the first four weeks of vegetative growths. However, during the same growth stages, both the Trykoside and *ComCat®* treatments contributed to increased plant diameter, although statistically non-significant. The Maxiflo treatment tended to have a slight but non-significant decreasing effect on plant diameter during the whole vegetative growth cycle. Six and eight weeks after planting, neither of the treatments influenced the plant's diameter significantly, but all treated plants tended to be smaller than the control plants. Furthermore, after the first two weeks, significant differences in plant height were observed in cabbage where treatment with Trykoside, both separately and in combination with Maxiflo, as well as the two commercial bio-stimulants *ComCat®* and Kelpak® tended to increase plant height compared to the Maxiflo treatment and untreated control.

The stem diameter of cabbage plants was significantly increased by the two bio-stimulants during the first four weeks after planting. Although the increase in stem diameter by these two bio-stimulants was also observed six and eight weeks after planting, the differences were not significant compared to the untreated control. Neither of the Maxiflo, Trykoside or Maxiflo/Trykoside combination treatments had a significant effect on stem diameter during the vegetative growth stage as measured over the first eight weeks after planting. Head and leaf fresh mass of cabbage compared to all other treatments except Trykoside, were also significantly improved by *ComCat®* treatment. Neither Maxiflo nor the Trykoside/Maxiflo combination treatment had a significant effect on either head or leaf mass. Both Maxiflo and Trykoside applied separately had a statistically significant decreasing effect on root fresh mass compared to the untreated control. However, although non-significantly, the opposite was observed when these two products were applied in combination. All other treatments had no significant effect on root mass.

Lettuce

In lettuce, the same tendency of all treatments to reduce plant height was observed at two-week intervals over the first eight weeks of vegetative growth compared to the control. However, this reduction was statistically significant only for the Trykoside/Maxiflo combination treatment while, overall, the Maxiflo treatment had the least reducing effect, followed by the two positive controls. Trykoside/Maxiflo combination treatment had the most severe reducing effect, and this was also statistically significant compared to the untreated control and most of the other treatments at least up to six weeks after planting. Statistically, no significant differences between treatments in terms of either head or leaf fresh mass were observed for lettuce. However, except for the Trykoside/Maxiflo combination treatment, all other treatments tended to have a decreasing effect on head fresh mass compared to the untreated control. Even though non-significant, the opposite tendency was observed for the combined Maxiflo/Trykoside treatment in terms of leaf mass. *ComCat*® also tended to decrease the leaf mass as compared to the control. Furthermore, Maxiflo and *Kelpak*® treatments had no effect on root development, whilst all other treatments significantly reduced the root fresh mass of lettuce plants.

Potato

Maxiflo and Trykoside treatments decreased the canopy area, which later became inconsistent, no clear pattern emerged, and no statistically significant effect was observed. The results also showed non-significant differences between the number of stolons and stem count per plant, probably due to huge standard deviations calculated. At the first count, four weeks after emergence, as well as at the third count, eight weeks after emergence, both Trykoside and Maxiflo seemed to have a slight enhancing effect on the number of stolons per plant. However, this was not the case at the second count. Furthermore, the lower control counts at four and eight weeks and the extremely low control counts after ten weeks, possibly indicate that more tubers have been formed from untreated seeds implicating the Maxiflo and Trykoside treatments to have had an inhibitory effect on tuber formation. Compared to the untreated control, especially Trykoside that had an increasing effect on the medium, medium/large and large size potato tubers. Maxiflo had no increasing effects on potato tuber sizes. However, there were indications that both Maxiflo and Trykoside treatments caused decreasing effects on the total tuber yield.

Tomato

In tomato, there were no significant differences between treatments in terms of either fruit size or total yield were observed. Except for the *Kelpak®* treatment that had no effect, all other treatments tended to reduce the total yield compared to the untreated control. This was especially visible for the large size fruits. Maxiflo and Trykoside did not have a significant effect on the total fruit yield; both tended to improve the number of medium and large size fruits during some of the harvest periods. This is not consistent with the findings of Kenny [60], who reported that the treatment of tomatoes with *Azospirillum* (the active ingredient of Maxiflo) significantly increased both the size and yield of tomato as well as green pepper fruits under field conditions. The author further observed a reduction in the occurrence of diseases in tomatoes. It was reported that *Azospirillum brasilense* increased the growth and yield of tomato plants.

A possible over-application of the products might have been a reason for the yield inhibition in tomatoes. In light of the results from this study and other reports, the possibility exists that the Maxiflo and Trykoside concentrations used in this study might not have been optimal and needs to be verified in a follow-up study. What is important to note is that the objectives for applying one or both of these products should be formulated beforehand. If the objective is to control infection by bacterial or fungal infection or both, the trial layout and parameters measured should be adapted in future research. In the latter instance, glasshouse trials are suggested as it will be easier to control. However, if the objective is to improve yields in crops, a series of concentrations should be tested, despite the current recommendations of the manufacturers, in order to ascertain the optimum.

Wheat

Maxiflo had no significant effects on root volume, while Trykoside applied separately and in combination with Maxiflo slightly, increased the root volume of wheat. *ComCat®* showed the same tendency to increase root volume slightly but, when applied in combination with both Maxiflo and Trykoside, it caused a sharp decrease in root volume. No statistically significant differences between treatments were observed in root fresh mass compared to the untreated control but, there was a significant difference detected in root dry mass. Both Maxiflo and Trykoside applied separately significantly increased the root dry mass but, when applied in combination, had no effect. Both *ComCat®* applied separately and the Maxiflo/Trykoside/ *ComCat®* combination reduced the root dry weight, though not significantly, compared to the untreated control as well as all other treatments. Twenty ears were selected randomly in plots, and replicated four times to calculate an acceptable average number of kernels per ear.

All treatments contributed to an increase in the number of wheat kernels per ear. Of these, the Trykoside treatment was the only one that increased the kernel number significantly. Although all treatments contributed to an increase in the kernel weight per ear compared to the untreated control, this was more marked when Maxiflo and Trykoside were applied separately. When *ComCat®* was added to both Maxiflo and Trykoside and tested in combination, the dry mass increase of kernels was more marked, but not significant, than where Maxiflo and Trykoside were tested in combination. Maxiflo and *ComCat®* applied separately as well as the Maxiflo-Trykoside-*ComCat®* combination treatment increased the total dry kernel yield slightly. Trykoside, on its own, contributed to a reduction in kernel yield of almost 50%. Probably due to large standard deviations between replicates, this was again regarded as non-significant by the statistical procedure.

FINAL CONSIDERATIONS

The application of microorganisms such as the fungus *Trichoderma* and the bacterium *Azospirillum* spp. is a well-established technique in organic farming systems. However, the natural products manufactured from these organisms in the past were mostly in powdered forms. A company Axiom Bio-Products Pty Ltd, recently established in South Africa, manufactured two products, namely Maxiflo *(Azospirillum* based) and Trykoside *(Trichoderrna* based) in liquid form. The effect of these two products on the growth and yield of *Pisum sativum* and other five economically important crops was investigated, strictly following the instructions of the manufacturers. The two commercially available natural products, *ComCat®* and *Kelpak®,* were also used as positive controls in order to ascertain the impact of Maxiflo and Trykoside, both separately and in combination, by comparison. Findings made from the experiment with peas, a legume crop, differed entirely from those observed with cabbage and lettuce.

Maxiflo applied separately and, together with Trykoside, had a significant effect on seed yield, while Trykoside on its own had no effects. Furthermore, *ComCat®,* a commercial bio-stimulant used as a positive control, was added to the Maxiflo-Trykoside mixture, significant enhancement in pod fresh mass, pod number and seed mass was observed. *ComCat®* is known to enhance flower formation in crops [6]. This would supply a simple explanation for the results obtained with the three-way application treatment. As the main enhancing effect of Maxiflo was on seed yield only, it might be indicative of a different mechanism of action than that of *ComCat®.* Additionally, a combination treatment might indicate a synergistic effect between products in the case of peas. Only a few reports on the effect of *Azospirillum* on the yield of legumes are available in the literature and none on the effect of *Trichoderma.*

Field inoculation of garden peas and chickpeas with *A. brasilense* produced a significant increase in seed yield but did not affect the dry matter yield [71]. The author suggested that the increase in yield was the result of increased nitrogen fixation after inoculation with *Azospirillum*. However, according to Steenhoudt & Vanderleyden [63] *Azospirillum spp.* secretes phytohormones that might explain the possible involvement of an "X-factor" in stimulating vegetative growth [68]. Bashan & Holguin [50] also reported on the growth-regulating properties of *Azospirillum* and the possible involvement of phytohormones secreted by the bacterium. However, no mention in literature has been made of a possible secreted chemical or hormone that might meet the description of an "X-factor" for *Trichoderma*. Traditionally, *Trichoderma* inoculants are applied to the soil. Zheng & Shetty [73] reported that soil sanitation with *Trichoderma viride, T. harzianum* and *T. pseudokoningii* increased the germination rate of pea seeds by 20, 40 and 15%, respectively, compared to seed pretreatment before sowing.

However, both *Azospirillum* and *Trichoderrna* have been known for their parasitic effect against soil-borne pathogenic organisms for at least two decades [72, 73]. Okon & Labandera-Gonzalez [29] reported that *Azospirillum* is an N_2-fixing rhizobacterium, which lives in close association with plants and is capable of increasing the yield of important crops grown in various soils and climatic regions. Significant yield increases in the order of 5 to 30% have been reported. This was attributed to improved root development and enhanced water and mineral uptake. Based on preliminary field results, the manufacturers of the liquid formulations of Maxiflo and Trykoside recently postulated that the liquidising process might have played a role in releasing an "X-factor" with biostimulatory properties from these organisms. For this reason, the emphasis was placed on foliar applications in this study in an attempt to authenticate this postulation, which so far proved positive.

CONCLUSION

Finally, in light of the observation that treating *Pisum sativum* L. with a combination of the two products Maxiflo and Trykoside led to enhanced growth or yield or both in some, while evoking an opposite response in other selected crops. These findings clearly indicate that follow-up studies involving different ratios of the products remain worthwhile to pursue. As already indicated, what is also important to note is that the objectives for applying one or both of these products should be formulated beforehand, and if the objective is to control infection by bacterial or fungal infection or both, the trial layout and parameters measured should be adapted in future research. As stated earlier, if the objective is to improve yields in legume crops, a series of concentrations and combination ratios should be tested in order to ascertain the optimum for each species and genotype. From this study, especially with respect to potatoes, it seems that the advantage of applying biological agents such as *Azospirillum* and *Trichoderma* also lies in its ability to control pathogens. From an organic farming perspective, this might be the way to go.

LIST OF ABBREVIATIONS

BNF	Biological nitrogen fixation
BRs	Brassinosteroids
GA₃	Gibberellic acid
ISR	Induced systemic resistance
K	Potassium
L	Litre
LSD	Least significant difference
N	Nitrogen
N/ha	Nitrogen per hectare
O₂	Oxygen
P	Phosphorus
PGRs	Plant growth regulators

CONSENT FOR PUBLICATION

Not applicable.

CONFLICT OF INTEREST

The author declares no conflict of interest, financial or otherwise.

ACKNOWLEDGEMENTS

Declared none.

REFERENCES

[1]   Qaim M, Kouser S. Genetically modified crops and food security. PLoS One 2013; 8(6): e64879. [http://dx.doi.org/10.1371/journal.pone.0064879] [PMID: 23755155]

[2]   De Vries PFWT. Food Security? We are losing ground fast! In: Nosberger J, Geiger HH, Struck PC, Eds. Crop science: Progress and prospects. GABI Publishing 2001; pp. 1-14.

[3]   Heidhues F. The future of world, national and household food security. In: Nosberger J, Geiger HH, Struik PC, Eds. Crop science: Progress and prospects. U. K.: CABI Publishing, Cromwell Press 2001; pp. 15-30.

[4]   Nelson R, Orrego R, Ortiz O, *et al.* 2001. Working with resource poor farmers to manage plant diseases. Plant Dis 2001; 85(7): 684-95.

[http://dx.doi.org/10.1094/PDIS.2001.85.7.684] [PMID: 30823190]

[5]   Roth U, Friebe A, Schnabl H. Resistance induction in plants by a brassinosteroid-containing extract of *Lychnis viscaria* L. Zeitschrift Fur Naturforschung 2000; 54: 1-25.
[http://dx.doi.org/10.1515/znc-2000-7-813]

[6]   Schnabl H, Roth U, Friebe A. Brassinosteroid-induced stress tolerances of plants. Rec Res Dev Phytochem 2001; 5: 169-83.

[7]   Putnam AR, Tang CS. The Science of allelopathy. New York: Wiley 1986; pp. 13-58.

[8]   Chou CH, Waller GR, Cheng CS, Yang CF, Kim D. Allelochemical activity of naturally occurring compounds from mung bean (*Vigna radiata* L.) plants and their surrounding soil. Bot Bull Acad Sin 1995; 36(1): 9-18.

[9]   Gorris LGM, Smid EJ. Potato of flower bulbs using antifungal plant metabolites. Brighton Crop Protection Conference. Pests and Diseases. 801-6.

[10]   Rizvi SJH, Rizvi V. Exploitation of allelochemicals in improving crop productivity Allelopathy: Basic and applied aspects. London: Chapman and Hall 1992; pp. 443-73.

[11]   Davies DR, Berry GJ, Heath MC, Dawkins TCK. Pea *(Pisum sativum* L.). In: Summerfield RJ, Roberts EH, Eds. Grain legume crops. London, UK: Williams Collins Sons and Co Ltd. 1985; pp. 147-98.

[12]   Slinkard AE, Bascur G, Hernandez-Bravo G. Biotic and abiotic stresses of cool season food legumes in western hemisphere. In: Muehlbauer FJ, Kaiser WJ, Eds. Expanding the production and use of cool season food legumes. Dordrecht, The Netherlands: Kluwer Academic Publishers 1994; pp. 195-203.
[http://dx.doi.org/10.1007/978-94-011-0798-3_10]

[13]   Duke JL. Handbook of legumes of world economic importance. New York: Plenum Press 1981; pp. 199-265.
[http://dx.doi.org/10.1007/978-1-4684-8151-8]

[14]   Singla R, Chahal SS, Kataria P. Economics of production of green peas (Pisum sativum L.) in Punjab. Agric Econ Res Rev 2006; 19: 237-50.

[15]   Agricultural statistics-2009-2010. Washington, D. C.: United State 2010.

[16]   Olivier OJ. Climatic requirements for cole crops. Vegetable and ornamental plant institute. Agricultural Research Council Roodeplaat 1995.

[17]   Simmonds NW. Evolution of crop plants. London: Longman 1976; p. 36.

[18]   Jackson DC. Climatic requirements for cole crops. Vegetable and ornamental plant institute. Agricultural Research Council Roodeplaat 1998.

[19]   Ryder EJ. Lettuce breeding (part of vegetable breeding). Westport, Conn: A.V.L. Publishing Co. 1986; pp. 436-72.

[20]   Robinson RW, McCreight JD, Ryder EJ. The genes of lettuce and closely related species. Plant Breed Rev 1983; 1: 267-94.

[21]   Jansen M. Vegetables crop profile: lettuce selected references and abstracts. Vegetable and potato producers. Kentville, NS: Association of NOVA Scotia, Kentville Agricultural Centre 1994; p. 163.

[22]   Cornell HJ, Hovelling AW. Wheat chemistry and utilisation. Lancaster: Technomic Publishing Company 1998; pp. 43-53.

[23]   Metcalf DD, Wilson CR. The process of antagonism of *Sclerotium cepivorum* in white rot affected onion roots by *Trichoderma koningii*. Plant Pathol 2001; 50: 249-57.
[http://dx.doi.org/10.1046/j.1365-3059.2001.00549.x]

[24]   Chet I. *Trichoderma-application,* mode of action and potential as a biocontrol agent of soil-borne plant pathogenic fungi. In: Chet I, Ed. Innovative approaches to plant disease control. New York, USA:

John Wiley and Sons 1987; pp. 137-67.

[25]   McCollum JP, Ware GW. Producing vegetable crops. Danville, USA: The Interstate Printers and Publishers INC 1975; p. 175.

[26]   Agwah EMR, Shahaby AF. Associative effect of *Azospirillum* on vitamin C, chlorophyll content and growth of lettuce under field conditions. Ann Agric Sci 1993; 38(2): 423-34.

[27]   Fallik E, Sarig S, Okon Y. Morphology and physiology of plants roots associated with *Azospirillum*. In: Okon Y, Ed. Azospirillum/plant associations. FL: CRC Press Boca Raton 1994; pp. 77-86.

[28]   Okon Y, Labandera-Gonzalez CA. Agronomic applications of *Azospirillum*- An evaluation of 20 years worldwide field inoculation. Soil Biol Biochem 1994; 26: 1591-601.
[http://dx.doi.org/10.1016/0038-0717(94)90311-5]

[29]   Wani SP. Inoculation with associative nitrogen fixing bacteria: Role in cereal grain production improvement. Indian J Microbiol 1990; 30: 363-93.

[30]   Roiger DJ, Jeffers SN, Caldwell RW. Occurrence of *Trichoderma* species in apple orchard and woodland soil. Soil Biol Biochem 1991; 23: 353-9.
[http://dx.doi.org/10.1016/0038-0717(91)90191-L]

[31]   Chet I. Biological control of soil-borne plant pathogens with fungal antagonists in combination with soil treatments. In: Hornby D, Ed. Biological control of soil-borne plant pathogens. Wallingford, UK: C.A.B.I. International 1990; pp. 15-25.

[32]   Kloepper JW. Plant growth-promoting rhizobacteria (other systems). In: Okon Y, Ed. Azospirillum/plant association. London, UK: CRC Press 1994; pp. 137-66.

[33]   Whipps JM, Lumsden RD. Biological control of *Pythium* species. Biocontrol Sci Technol 1991; 1(2): 75-90.
[http://dx.doi.org/10.1080/09583159109355188]

[34]   Wilson CL, Winsiewski ME, Biles CI, McLaughlin R, Chalutz E, Droby S. Biological control of post-harvest diseases of fruits and vegetables: alternatives to synthetic fungicides. Crop Prot 1991; 10: 172-6.
[http://dx.doi.org/10.1016/0261-2194(91)90039-T]

[35]   Elad Y, Chet I. Practical approaches for biocontrol implementation. In: Rreuveni R, Ed. Novel approaches to integrated pest management. Boca Raton, FL, USA: CRC Press 1995; pp. 323-38.

[36]   Smith VL, Wilcox WF, Harman GE. Potential for biological control of Phytophthora root and crown rots of apple by *Trichoderma* and *Gliociadium* spp. Phytopathology 1990; 80: 880-5.
[http://dx.doi.org/10.1094/Phyto-80-880]

[37]   Sivan A, Elad Y, Chet I. Biological control effects of a new isolate of *Trichoderma harzianum* on *Pythium aphanidermatum*. Phytophol 1984; 74: 498-501.
[http://dx.doi.org/10.1094/Phyto-74-498]

[38]   Marois JJ, Johnston SA, Dunn MT, Papavizas GC. Biological control of verticillum wilts of eggplant in the field. Plant Dis 1982; 66: 1166-8.
[http://dx.doi.org/10.1094/PD-66-1166]

[39]   Knudsen GR, Eschen DJ, Dandurand LM, Bin L. Potential for biocontrol of *Sclerotina sclerotiorum* through colonization of *Sclerotia* by *Trichoderma harzianum*. Plant Dis 1991; 5: 466-70.
[http://dx.doi.org/10.1094/PD-75-0466]

[40]   Yedidia I, Benhamou N, Chet I. Induction of defense responses in cucumber plants *(Cucumis sativus* L.) By the biocontrol agent *Trichoderma harzianum*. Appl Environ Microbiol 1999; 65(3): 1061-70.
[http://dx.doi.org/10.1128/AEM.65.3.1061-1070.1999] [PMID: 10049864]

[41]   Tugarova AV, Vetchinkina EP, Loshchinina EA, Burov AM, Nikitina VE, Kamnev AA. Reduction of selenite by *Azospirillum brasilense* with the formation of selenium nanoparticles. Microb Ecol 2014; 68(3): 495-503.

[http://dx.doi.org/10.1007/s00248-014-0429-y] [PMID: 24863127]

[42]   Blasczyk L, Siwulski M, Sobieralski K, Lisiecka J, Jedryczka M. *Trichoderma* spp.– application and prospects for use in organic farming and industry. J Plant Prot Res 2014; 54(4): 309-16.
[http://dx.doi.org/10.2478/jppr-2014-0047]

[43]   Keinath AP, Fravel DR, Papavizas GC. Evaluation of potential antagonists for biocontrol of *Verticillium dahlia*, In Procedures of the 5th International Verticillum Symposium., Leningrad, USSR, 82.

[44]   Sharon E, Bar-Eyal M, Chet I, Herrera-Estrella A, Kleifeld O, Spiegel Y. Biological control of the root-knot nematode *Meloidogyne javanica* by *Trichoderma harzianum*. Phytopathology 2001; 91(7): 687-93.
[http://dx.doi.org/10.1094/PHYTO.2001.91.7.687] [PMID: 18942999]

[45]   Baker R, Elad Y, Chet I. The controlled experiment in the scientific method with special emphasis on biological control. Phytopathol 1984; 74: 1-1019.
[http://dx.doi.org/10.1094/Phyto-74-1019]

[46]   Jing GLI, Benoit F, Ceustermans N. Studies on effect of different nutrient solution treatments on disease control and yield improvement of lettuce in NFT system. Institute of Horticulture, Zhejiang Academy of Agricultural Sciences, Hangzhou, Zhejiang. 2001; 31002: pp. (5)13-6.

[47]   Harman GE. Myths and dogmas of biocontrol: Changes in perceptions derived from research on *Trichoderma harzianum* T-22. Plant Dis 2000; 84(4): 377-93.
[http://dx.doi.org/10.1094/PDIS.2000.84.4.377] [PMID: 30841158]

[48]   Howell CR, De Vay JE, Garber RH, Batson WE. Field control of cotton seedling diseases with *Trichoderma virens* in combination with fungicide seed treatments. J Cotton Sci 1997; 1: 15-20.

[49]   Howell CR, Hanson LE, Stipanovic RD, Puckhaber LS. Induction of terpenoid synthesis in cotton roots and control of *Rhizoctonia solani* by seed treatment with *Trichoderma virens*. Phytopathology 2000; 90(3): 248-52.
[http://dx.doi.org/10.1094/PHYTO.2000.90.3.248] [PMID: 18944616]

[50]   Bashan Y, Holguin G. *Azospirillum* plant relationships: environmental and physiological advances (1990-1996). Can J Microbiol 1997; 43: 103-21.
[http://dx.doi.org/10.1139/m97-015]

[51]   Zuberer DA. Biological dinitrogen fixation: introduction and non-symbiotic. In: Sylvia DM, Fuhrmann JJ, Hartel PG, Zuberer DA, Eds. Principles and applications of soil microbiology. New Jersey: Prentice Hall 1998; pp. 295-321.

[52]   Gaskins MH, Hubbell DH. Response of non-leguminous plants to root inoculation with free-living diazotrophic bacteria. In: Harley JL, Russell RS, Eds. The soil root interface. London, UK: Academic Press 1981; pp. 175-82.

[53]   Van de Broek A, Vanderleyden J. Review: Genetics of the *Azospirillum* plant root association. Crit Rev Plant Sci 1995; 14: 445-66.
[http://dx.doi.org/10.1080/07352689509701932]

[54]   Okon Y, Itzigsohn R. The development of *Azospirillum* as a commercial inoculant for improving crop yields. Biotechnol Adv 1995; 13(3): 415-24.
[http://dx.doi.org/10.1016/0734-9750(95)02004-M] [PMID: 14536095]

[55]   Gadagi R. Studies on *Azospirillum* isolates of ornamental plants and their effect on *Gaillardia pulchella* var picta fouger. University of Agricultural Sciences Dharwad, India, New Delhi, PhD Thesis 1999.

[56]   James EK. Nitrogen fixation in endophytic and associative symbiosis. Field Crops Res 2000; 65: 197-209.
[http://dx.doi.org/10.1016/S0378-4290(99)00087-8]

[57]   Fritzsche C, Ueckert J, Niemann EG, Polsinelli M, Materassi R, Vincenzini M. Growth parameters of microaerobic diazotrophic rhizobacteria determined in continuous culture. Nitrogen fixation. Proceedings of the 5th International Symposium on Nitrogen fixation with Non-Legumes, Florence, Italy. Dev Plant Soil Sci. 48(): 231-4.

[58]   Holguin G, Patten CL, Glick BR. Genetics and molecular biology of *Azospirillum*. Biol Fertil Soils 1999; 29: 10-23.
[http://dx.doi.org/10.1007/s003740050519]

[59]   Bashan Y, Levanony H. Current status of *Azospirillum* inoculation technology: *Azospirillum* as a challenge for agriculture. Can J Microbiol 1990; 36(9): 591-608.
[http://dx.doi.org/10.1139/m90-105]

[60]   Kenny L. Survey on Mediterranean organic agriculture. Country Report, Morocco 2001.

[61]   El-Nagger AI, Mahamoud SM. Effects of inoculation with certain *Azospirillum* strains and nitrogen fertilizers on *Narcissus tazetta* L. under different soil texture. Assuit J Agric Sci 1994; 25(14): 135-51.

[62]   Steenhoudt O, Vanderleyden J. *Azospirillum,* a free-living nitrogen-fixing bacterium closely associated with grasses: genetic, biochemical and ecological aspects. FEMS Microbiol Rev 2000; 24(4): 487-506.
[http://dx.doi.org/10.1111/j.1574-6976.2000.tb00552.x] [PMID: 10978548]

[63]   Leclerc Y, Dannelly DJ, Seabrook JEA. Micro-tuberization of layered shoots and nodal cuttings of potato: The influence of growth regulators and incubation periods. Plant Cell Tissue Organ Cult 1994; 37: 113-20.
[http://dx.doi.org/10.1007/BF00043604]

[64]   Forti E, Mandolino G, Ranalli P. 1991. *In vitro* tuber induction influence of the variety and of the media. Acta Hortic 1991; 300: 127-31.

[65]   Barassi CA, Sueldo RJ, Alvarez MI. Effect of *Azospirillum* on coleoptile growth in wheat seedlings under water stress. Cereal Res Commun 1996; 24(1): 101-7.

[66]   Cooke HJ. Evolution of the long range structure of satellite DNAs in the genus Apodemus. J Mol Biol 1975; 94(1): 87-99.
[http://dx.doi.org/10.1016/0022-2836(75)90406-4] [PMID: 1142437]

[67]   Burdman S, Jurkevitch E, Schwartsburd B, Hampel M, Okon Y. Aggregation in *Azospirillum brasilense*: effects of chemical and physical factors and involvement of extracellular components. Microbiology (Reading) 1998; 144(Pt 7): 1989-99.
[http://dx.doi.org/10.1099/00221287-144-7-1989] [PMID: 9695932]

[68]   Molahlehi L. Chemical factors influencing dry bean yield. University of the Free State, Bloemfontein, Republic of South Africa, MSc Dissertation 2000.

[69]   Arteca RN. Brassinosteroids. In: Davies PJ, Ed. Plant hormones: Physiology, biochemistry and molecular biology. London: Kluwer Academic Publishers 1995; pp. 206-13.

[70]   Anwar A, Liu Y, Dong R, Bai L, Yu X, Li Y. The physiological and molecular mechanism of brassinosteroid in response to stress: a review. Biol Res 2018; 51(1): 46.
[http://dx.doi.org/10.1186/s40659-018-0195-2] [PMID: 30419959]

[71]   Papavizas GC. Biological control in crop production. NJ Totowa: Allanheld, Osmun, 1981; 461.

[72]   Cook RJ, Baker KF. The nature and practice of biological control of plant pathogens. Amer Phytopathol Soc 1983; p. 539.

[73]   Zheng ZX, Shetty K. Effect of apple pomance-based *Trichoderma* inoculants on seedling vigour in pea *(Pisum sativum)* germinated in potting soil. Process Biochem 1999; 34(6-7): 731-5.
[http://dx.doi.org/10.1016/S0032-9592(98)00149-6]

CHAPTER 5

Effect of EMS-Based Mutagenesis on Growth and Yield Response in Tepary Bean (*Phaseolus acutifolius*)

Andries Thangwana[1,*] and Phetole Mangena[2]

[1] *Irrigation and Climate Control Department (ICC), Flamingo Horticulture, Plot 25 Delarey Farm, Syferbult Road, Tarlton 1749, South Africa*

[2] *Department of Biodiversity, School of Molecular and Life Sciences, Faculty of Science and Agriculture, University of Limpopo, Private Bag X1106, Sovenga, 0727, South Africa*

Abstract: Tepary bean is an important food legume cultivated in semi-arid areas in many parts of the world, including sub-Saharan Africa. This crop is highly tolerant to drought and pests, but it is also generally low yielding. Similar to many other legumes, the genetic improvement of the tepary bean is also limited by its narrow genetic base. Techniques, such as mutation breeding, particularly chemical mutagenesis with ethyl methanesulfonate (EMS) have been successfully tested to induce genetic variability in this crop. As part of the few reports of chemical mutagenesis in tepary bean, this paper evaluated the seedling performance of the M_1 EMS mutagenised plants and adult plant performance of the subsequent generations of tepary bean under different conditions. Based on the results, EMS induced some dominant mutations that were detectable in the M_1 generation. These effects were further extended to the early mutagenised generations (M_2 to M_4) of the tepary bean under field conditions, revealing novel information regarding the response of tepary bean to chemical mutagenesis at both the seedling and adult plant stages.

Keywords: Ethyl methane sulfonate, growth, mutagenesis, tepary bean, yields.

INTRODUCTION

Tepary bean (*Phaseolus acutifolius*) is a short duration crop belonging to the bean family (Fabaceae). It is an edible, small-seeded, annual legume similar to dry bean (*Phaseolus vulgaris*). Tepary bean is indigenous to the south-western United States of America and Mexico [1].

[*] **Corresponding author Andries Thangwana**: Irrigation and Climate Control Department (ICC), Flamingo Horticulture, Plot 25 Delarey Farm, Syferbult Road, Tarlton 1749, South Africa; Tel: 2710 591 1040 & 2715 268 4715; E-mails: thangs1981@gmail.com & andries.thangwana@flamingo.net

Mohammed *et al*. [2] and Thomas *et al*. [3] reported that tepary bean is highly tolerant to drought, heat stress [4] and it is also resistant to many fungal diseases [5]. Tepary bean is produced mainly for human consumption. The pod usually contains 5–6 small grains and the fresh green pods are about 5.0–7.0 cm in length. Tepary bean seeds are consumed dry in different ways, including boiling, steaming, frying, or baking. Dry seeds can also be mixed with whole-grain maize or soups to make a high protein meal. In most cases, it is consumed as fresh green beans or as bean sprouts. Many studies showed that the tepary bean is superior to other bean types in the number of proteins, calcium, iron, magnesium, zinc, phosphorous and potassium [6, 3].

The dry seeds of tepary bean contain about 24.0 to 27.0% of proteins as compared to roughly 21.0% total in legumes such as common bean and faba bean. The seeds contain proteins, dietary fibre and vitamins that are highly suitable for human consumption [7]. Amongst other legumes, tepary bean seeds have been proven to contain high lectin activity after soybean. In addition, the seeds contain protease inhibitors that inhibit the growth of certain types of cancer cells. The lectin toxins present in tepary bean were reported by De Mejia *et al*. [8] to be potentially useful in chemotherapy for treating cancer. Furthermore, tepary bean remains traditionally ideal for health promotion in people suffering from the different types of diabetes [9]. However, Bhardwaj *et al*. [7] indicated that the protein content and mineral compositions in tepary bean make it ideal for full use as a food and feed crop, in addition to the natural antioxidants which contribute to the reduction of many cardiovascular diseases.

Miklas *et al*. [6] reported that the grain yield of tepary bean is generally low, partly because of the poor seed quality. The genetic base of tepary bean is also very narrow, but, potentially, it could be broadened by induced mutagenesis as accomplished in other legumes such as cowpea, lentil, mung bean and soybean using the mutagen ethyl methanesulfonate (EMS) [10 - 13]. Currently, there are very few reports in the literature on the induction of tepary bean mutants using EMS. Generally, the production of tepary bean, particularly in South Africa, is also constrained by the lack of improved commercial cultivars. Smallholder farmers and home growers cultivate traditional varieties largely in mixed cropping systems. Such practices often involve retaining seeds from each harvest for the next crop season. At times, the seeds are exchanged between growers; thus, increasing the possibility of seed mixing during the exchange process. This legume crop continues to provide food and nutrition security for rural people and resource-constrained farmers in sub-Saharan Africa (SSA) [14].

ORIGIN AND DISTRIBUTION OF TEPARY BEAN

Phaseolus acutifolius is a diploid (2n=2x=22) and self-pollinating annual leguminous plant, native to Arizona, United States and Mexico (Figs **1** and **2**) [1, 3]. It is a short-duration summer crop grown in various parts of the world, including Africa, and some parts of Europe and South Asia [15]. Currently, tepary bean is cultivated in some few African countries such as Kenya, Botswana and South Africa [16, 17]. American agriculture traditionally relied upon tepary bean for thousands of years for food and other services. Upon domestication of the variety *P. acutifolius* var. *acutifolius*, together with other legume species like chickpea, field pea, common bean, lentil, dry bean and soybean, all these crops have increased the value of legumes worldwide. This domesticated variety is bushier and well adapted to a very hot and dry environment [18]. An example of a densely sturdy tepary bean stand with a canopy stand is illustrated in Fig. (**1**) below. According to Gepts [19], the genus contains more than 70 wild plant species, distributed exclusively in the Americas, with a clear focal point in Mexico and Central America (Fig. **2**). These regions comprise the largest number of species found within the *Phaseoleae* tribe, with very prominent species such as *Phaseolus vulgaris* that are widely utilised in the SSA region and India.

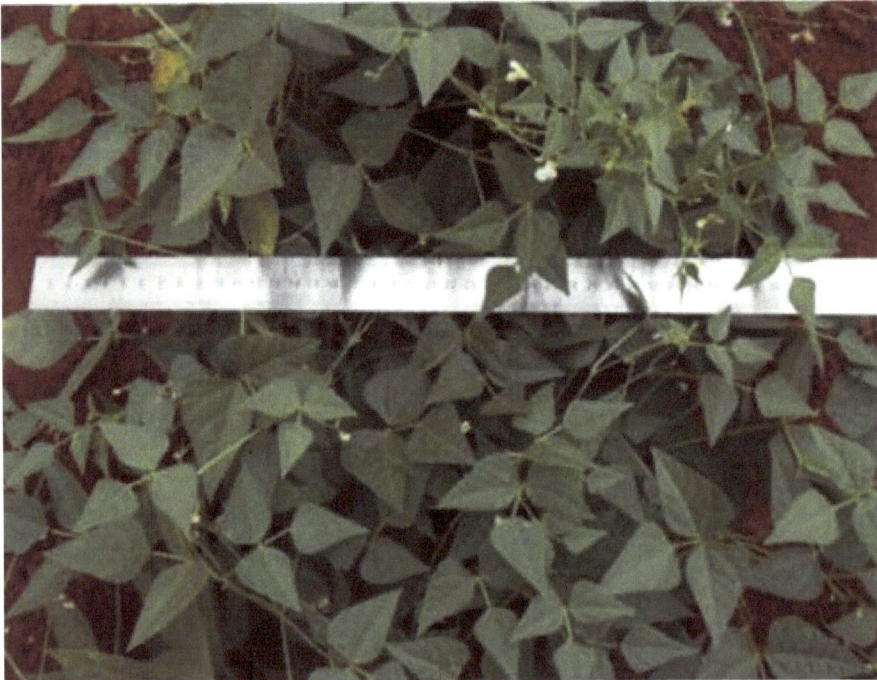

Fig. (1). Lateral section of a bushy tepary bean stand showing a canopy width measured using a metric ruler (42.5 cm).

Tepary bean remains one of the underutilised species, together with *Canavalia maritima, Psophocarpus tetragonolobus, Tylosema esculentum, Vigna umbrellata* and *Vigna subterranean*. Based on the analysis of geographical information system reported by Cortes *et al.* [20], plant species of this genus (*Phaseolus*) were distributed among the different precipitation regimes, especially in areas where drought is severe, such as the drier parts of Africa, north-eastern Brazil, coastal Peru, and northern highlands of Mexico.

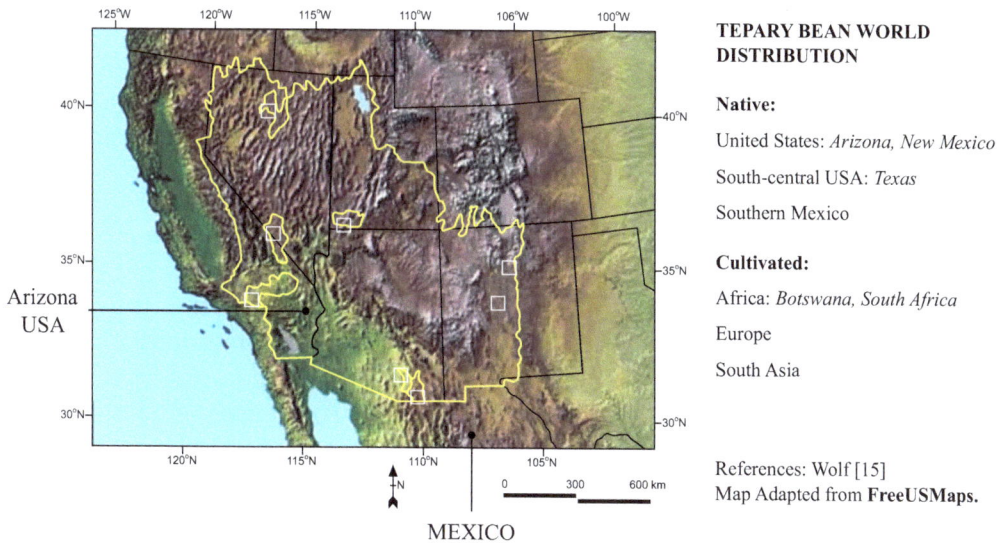

TEPARY BEAN WORLD DISTRIBUTION

Native:

United States: *Arizona, New Mexico*

South-central USA: *Texas*

Southern Mexico

Cultivated:

Africa: *Botswana, South Africa*

Europe

South Asia

References: Wolf [15]
Map Adapted from **FreeUSMaps.**

Fig. (2). Map highlighting the origin of tepary bean (*Phaseolus acutifolius*) and its areas of distribution worldwide.

BIOLOGY OF TEPARY BEAN

Tepary bean is easily distinguished from other bean species by its epigeal germination (Fig. **3A**) and grows relatively fast. It can be climbing or trailing with stems growing up to 4.0 m long. The leaves are trifoliate with narrow pointed end leaflets. The crop can still grow well under extreme environmental conditions. The seed absorbs water easily in moist soils leading to quick germination, particularly in white-seeded types [22]. The seed of domesticated types have no dormancy, and this serves as a disadvantage in humid regions, where seeds from shattered pods can germinate rapidly. The rate of germination increases with increasing temperatures from 12°C to 31°C [23]. The taproot, which is also a major diagnostic characteristic of all legumes, enables the plant to absorb moisture from deep soil layers that are not accessible by other crops with shallow root systems thus, enhancing the physiological capacity for water and nutrient uptake [24].

The moisture absorption capacity through the root system can also be influenced by the root biomass, particularly the lateral secondary and tertiary roots. In a study conducted under greenhouse conditions, profuse branching in the root system of tepary bean improved access to water. In cowpea, root length density declined significantly under moisture stress in cultivars that were sensitive to drought. These findings in tepary bean and cowpea studies were reported by Butare *et al.* [25] and Matsui and Singh [26], respectively. Compared to other legumes, flowering in tepary beans occurs within 27–40 days, generally maturing early. In the tropics, short-duration types may mature within two months, but most wild and domesticated varieties require up to 68 days in order to reach maturity. This suggests that multiple varieties of crops of tepary bean are suitable for planting in a given season; thus, securing food availability. Furthermore, this crop has an advantage as a short-duration crop in smallholder production systems, particularly, in areas that are prone to terminal drought stress [4].

The mature pods are 4.5–6.5 cm long, containing 2 to 7 small grains. At maturity, the pods are prone to shattering. In order to avoid pod shattering, the pods are usually harvested by hand as soon as they turn brown. Sometimes, the whole plants are pulled up by hand for threshing, which normally requires drying of fruit pods for a few days before they are threshed [21]. Apart from pod shuttering, another limitation to commercial production of the crop is the lack of uniformity in the maturing of fruit pods. Therefore, improved cultivars of tepary bean that are resistant to pod shattering and mature uniformly could enhance mechanized commercial production of the crop [27].

THE IMPORTANCE OF TEPARY BEAN

Leguminous crops are important in animal and human nutrition, particularly for the larger amounts of dietary fibre, proteins, carbohydrates, oils, vitamins, essential mineral nutrients (calcium, iron, magnesium, zinc, phosphorous and potassium) and various essential amino acids (methionine, cysteine, leucine, lysine, arginine, histidine and phenylalanine) [28]. Tepary bean also covers dietary protein and energy requirements for humans and provides forage or hay for livestock feeds [29, 30]. Potential value and multiple benefits that could emanate from this crop include the fact that it has showed resistance to various fungal diseases such as bean rust, common bacterial blight and bean golden mosaic virus when compared to other legumes like cowpea [31 - 33]. It is also highly tolerant of saline soil than common bean [34]. Further to these, it showed considerable resistance against insect pests such as leafhoppers (*Empoasca* spp.) and bruchids (*Acanthoscelides obtectus*) [35 - 37].

Tepary bean was used in wide crosses of introgressive hybridisation of useful genes into common bean (*Phaseolus vulgaris*) due to the abovementioned desirable traits [38 - 40]. The use of this crop as breeding germplasm for new common bean lines is highly motivated by the need to develop new cultivars that achieve increased yield potential even under abiotic and biotic stress conditions [41]. The inherent desirable characteristics that the tepary bean possesses make it an ideal donor crop for convergent gene pairs that control key developmental and stress resistance traits, especially for cultivation in various parts of sub-Saharan Africa [42]. According to Rao *et al.* [4], Mohammed *et al.* [2] and Federici *et al.* [43], the ability of tepary bean to tolerate drought makes it useful in cropping systems that are prone to moisture stress. The drought tolerance in this crop is attributed partly to sensitive stomata, which can close at relatively high-water potentials [44, 45].

Tepary bean has a deep and extensive root system that can extract soil moisture, which cannot be accessible to other competitive herbaceous weeds and legumes [46] and contributes to the improvement of soil fertility through biological nitrogen fixation (BNF) [17, 47]. This process of biological fixation of N enhances soil fertility by increase soil nitrogen levels and organic matter of the soil, including attaining the desirable physical soil properties (tilth) such as aeration, structure, drainage, and water holding capacity. The crop is important in smallholder cropping systems particularly where inorganic nitrogenous fertilizers may be unavailable. Therefore, BNF remains a less expensive and sustainable form of nitrogen than the synthetic N fertilisers. It lessens the relatively large amounts of renewable or diesel gas energy inputs associated with synthetic N fertilisation of crops while reducing air and gas pollution [48].

MUTATION BREEDING

Mutation breeding is the process of treating germplasm with mutagens in order to generate mutants with useful traits that can be selected for cultivar development. The genetic variation achieved through this approach is generally random since any part of the genome can be mutated. In some cases, whole chromosome segments can be affected [49]. However, re-joining of chromosomes can result in useful gene combinations which can be exploited by plant breeders to identify desirable genotypes through systematic evaluation either at the whole plant or DNA level. Where DNA markers for identifying such useful mutations are absent, plant breeders resort to evaluating the phenotypes in the field (or greenhouse) in order to select desirable plant genotypes.

Types and Mechanisms of Mutagens

Mutagenic agents such as physical (gamma radiation, X-rays and UV light) or chemical mutagens (colchicine, ethylamine, EMS, diethyl sulfate and NaM_3, *etc.*) can be used to induce mutations [50 - 52]. Among the chemical mutagens, ethyl methanesulfonate (EMS) was widely used (Table **1**) and highly effective in inducing useful mutations in a wide range of crop species [10]. However, it is always necessary to determine the optimum dose for inducing useful mutations for a crop species of interest since these chemicals can be very lethal. A study by Ndou *et al.* [53] observed that a high dose of EMS at about 0.7% (*v/v*) drastically reduced seed germination and seedling growth in wheat. In cowpea, percent germination of M1 seeds was used for determining the lethal dose for 50.0% of the test material (LD50), as reported by Horn and Shimelis [54]. Lukanda *et al.* [55] and Mba *et al.* [50] reported significant differential responses to gamma radiation among the cowpea genotypes as measured by the epicotyl and hypocotyl lengths. The LD50 was regarded as an important parameter for measuring acute toxicity as well as for estimating the optimum dose.

Table 1. Examples of some legume and non-legume crop species in which mutagens were applied.

Mutagenic Agent	Species	Reference / Sources
Ethyl Methane Sulfonate	Cowpea (*Vigna unguiculata*)	Gnanamurthy and Dhanavel [56]
	Lentil (*Lens culinaris*)	Gaikwad and Kothekar [11]
	Wheat (*Triticum aestivum*)	Jamil and Khan [61]
	Mung bean (*Vigna radiata*)	Khan and Goyal [58]
	Oats (*Avena sativa*)	Verhoeven *et al.* [60]
X-rays	Wheat (*Triticum aestivum*)	Asghari *et al.* [64]
Gamma rays	Cowpea (*Vigna unguiculata*)	Horn and Shimelis [54]
	Soybean (*Glycine max*)	Celik *et al.* [63],
	Lentil (*Lens culinaris*)	Muhammed *et al.* [62]
	Mung bean (*Vigna radiata*)	Tah [57],
Colchicine	Soybean (*Glycine max*)	Mangena [52]
	Cowpea (*Vigna unguiculata*)	Essel *et al.* [59]
	Pigeon pea (*Cajanus cajan*)	Mangena [52], Essel *et al.* [59]

As also indicated in this study, the highest EMS dose (2.0% *v/v*) generally depressed the seedling vigour as indicated by all the attributes as they appear in Figs. (**3B** to **E**) with the exception of shoot dry weight (SDW) and root dry weight (RDW) (Table **2**). Nonetheless, shoot height (SHT) varied with EMS dose within the genotypes, but in some cases, the seedlings did not develop any roots and remained stunted. These findings suggested some new insights into the response of tepary bean to chemical mutagenesis using EMS. To a large extent, the results were consistent with the findings reported for the other field crops that were treated with various mutagenic agents. For instance, in cowpea and wheat, as reported by Ndou *et al.* [53] and Horn and Shimelis [54]. The reduced seed germination was previously attributed to water potential difference between the outside and the inside of the seed caused by higher EMS concentration leading to insufficient water absorption for proper germination [12].

Fig. (3). Morphology of tepary bean seedlings. Epigeal germination (**A**), effects of EMS on germinated seeds (**B** to **E**) and areal/side view of mutagenized seedlings evaluated for the effect of EMS on root and shoot development (**F** and **G**).

Till *et al.* [65] and Okagaki *et al.* [66] indicated that chemical mutagens work mostly by inducing point mutations in the nuclear DNA. According to Henry *et al.* [67], chemical mutagenesis efficiently generates phenotypic variation in otherwise homogeneous genetic backgrounds. The point mutations occur when a single base pair of a gene is changed and is almost exclusively guanine G/C to A/T conversions that occur randomly along the genome [68]. Point mutations can alter the coding sequence of amino acid, thus resulting in potentially altered protein structures. Changes in which a purine is converted to purine (A-G or G-A) or in which a pyrimidine is converted to pyrimidine (T-C or C-T) are called transitions and those in which a purine is converted to pyrimidine (A-C) or vice versa are termed transversions [69]. In many cases, EMS induces –C to –T changes resulting in C/G to T/A substitution [68, 70].

MORPHOLOGICAL AND YIELD TRAITS

In tepary bean, grain yield is of major economic importance. Consequently, it is necessary to evaluate yield attributes that influence grain yield in breeding programs aimed at developing improved cultivars of the crop. These attributes include the duration of flowering, morphology and physiological maturity, number of pods per plant, grain size as well as resistance to pests and diseases. In addition, the lack of improved cultivars partly limits the adoption of the crop on a commercial scale in many countries worldwide [19, 27]. In breeding programmes, genetic variation is a prerequisite for successful crop improvement. Mutation breeding, which is capable of creating useful genetic variability, has been successfully used for decades in the development of improved cultivars in several crops, including cereals and other grain legumes (Table 1). These techniques have proven useful in obtaining new useful traits and creating new genetic variability and combinations, as supported by Sangsiri *et al.* [71] and Anbarasan *et al.* [72].

Table 2. Analysis of variance for seven attributes of seedling vigour in M_1 seedlings among three tepary bean genotypes.

Source	Mean Square						
	G%	NSR	PRL (mm)	SRL (mm)	SHT (mm)	SDW (mg plant⁻¹)	RDW (mg plant⁻¹)
Rep	987.76*	15.8735	91.97	16.81	49.71	1.41	1.31*
Genotype	274.53	44.2002	199.97	85.42	69.33	0.61	1.90**
Rep x Genotype	91.5242	52.3545	470.72	369.57	184.85	2.20	0.20
Dose	1679.59**	712.34**	3840.62**	4997.00**	4925.6**	9.21*	7.31**
Genotype x Dose	92.04	186.10*	1600.72**	144.16	517.95	2.10	1.00**
Mean	77.56	20.58	52.75	28.87	53.93	0.07	0.0068
R² (%)	74.74	74.66	72.94	86.58	74.48	48.92	86.78
C.V. (%)	17.05	41.56	41.24	42.05	35.76	23.02	25.27

** Significant at the 1.0% probability level; * Significant at the 5.0% probability level. [%G = percent seed germination; NSR = number of secondary roots; PRL = primary root length; SRL = secondary root length; SHT = shoot height; RDW = root dry weight; SDW = shoot dry weight].

Effects of EMS on Growth and Yield of Tepary Bean

The findings previously made in our studies showed a variation in the effects of EMS, ranging from germination rates, seedling development (Fig. **3A – G**) and up to matured fruit-bearing tepary bean plants, based on morphology. In addition, the findings further showed a significantly high genotype x dose interaction for shoot height (SHT), shoot dry weight (SDW), primary root length (PRL) and root dry weight (RDW) in Table **2** and Fig. (**4**). The mean length of the primary roots among the seedlings was even greater than 40% of the secondary root length (SRL), while the SDW was at least 10-fold more than the DRW. The evaluations of percentage germination (Table **2**) also revealed EMS sensitivity that could be linked to the genotype as previously reported by Horn and Shimelis [54] for cowpea in which differential response to gamma radiation was observed among the genotypes. However, by interpolation, it would be reasonable to assume that EMS doses over 10% (*v/v*) could stop seed germination completely and may need to be avoided in this specific genotype.

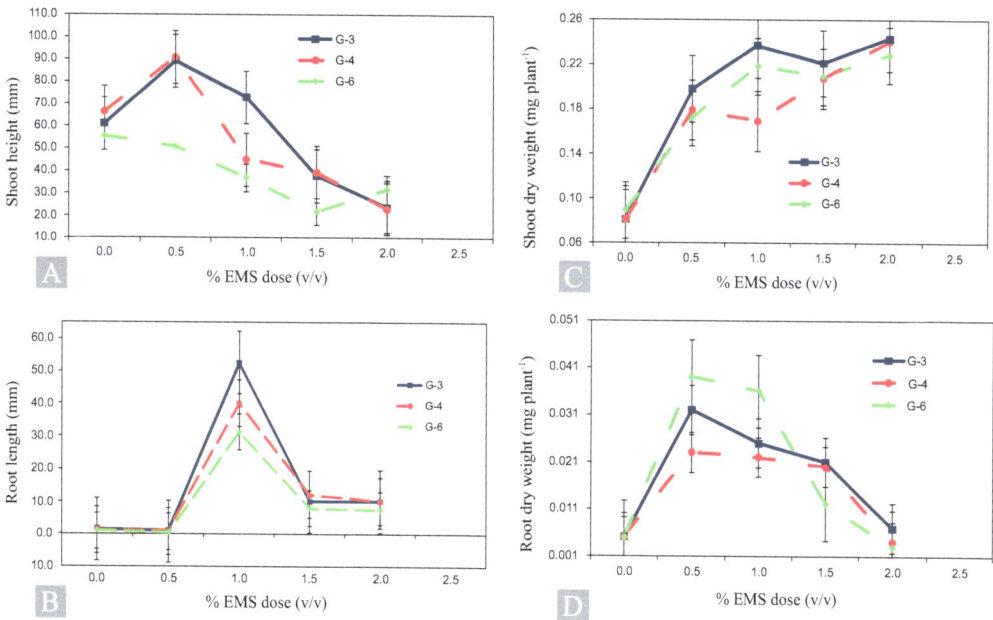

Fig. (4). Responses of tepary bean following pretreatment of seeds with different concentrations of EMS on shoot height and root length, as well as shoot and root dry weight.

The induced variations on mutagenized seedlings through the use of different EMS doses emphasise the potential that this technique has, especially to accomplish genetic changes that may be impossible to achieve *via* conventional breeding. Gnanamurthy and Dhanavel [56] reported induced morphological mutants and subsequent chromosomal variations in cowpea following treatment of dry seeds, variety CO-7, with different doses of EMS mutagen. Similarly, Girija and Dhanavel [10] also reported high-frequency chlorophyll and viable mutants in the M2 generation using the same CO-7 cowpea variety with 15 mM EMS. Further research continues to show that these changes are now being developed and characterised from seedling mutants to succeeding generations of matured plants. The findings made in this study and various others clearly provide comprehensive insights on mutants regeneration and the possibilities available to researchers today in the field of molecular breeding to complement the already existing inefficient breeding tools.

The Influence of EMS on Yield

In grain legumes, traits such as the deep root system, biomass accumulation, photosynthates partitioning and seed filling have been considered critical to yield improvement and seed quality (Fig. **4**). As reported by Mwale *et al.* [14], reproductive traits that include pod partitioning index, harvest and pod harvest index are also targeted for selection to improve yield gains, particularly under stress conditions. Although, great efforts have been undertaken to create viable mutants with genetic parameters for improved yield and yield contributing characters, as presented in Fig. (**5**). Plant selection for resistance to biotic and abiotic stress, combined with better yield attributes are still investigated. Currently, the different doses of EMS were found to be very effective in generating mutation for higher yield in various legumes. Even though Girija and Dhanavel [10] demonstrated a decrease in percentage seed germination with an increase in mutagen concentration, pod mutants showed higher yields with an optimum amount of EMS, especially in combination with gamma rays. This optimum amount of EMS induced 44.01% survival reduction with 3.26% mutation frequency [10].

Kozgar *et al.* [72] induced genetic variability in M_1 and M_2 generations of *Vigna radiata* and *Vigna mungo*, and found a linear correlation of the total plant yield following the genetic mutation using EMS to seed protein content and nitrate reductase activity. Khan *et al.* [74] reported sodium azide dose-dependent biological effects in *Vigna radiata* L., variety K-851 and Pusa Baisakhi. The SA is another type of mutagen widely reported to induce genomic and phenotypic mutations in many pulse crops. As observed in the use of EMS in Fig. **5** in this study, NaM_3 pretreatment of mung bean seeds affected yield by initially exhibiting inhibitions in seed germination and a further reduction in pollen fertility as well as survival maturity only in the M_1 plants generation. Such high degree chemically induced mutagenic damages were also reported in *Arabidopsis thaliana* [68] and japonica rice (*Oryza sativa* L.) [75] using EMS mutagen.

Fig. (5). Example of Individual mutant plants segregated for yield analysis based on the number of pods in the subsequent generation, after M1 plant regeneration.

USE OF MUTAGENESIS FOR DEVELOPMENT OF STRESS TOLERANCE CULTIVARS

Grain legume species, like the rest of cereal crops, have a very limited genetic diversity to meet the growing demand to feed the ever-increasing global population. The poor genetic variability for growth and agronomic traits always serve as setbacks for the genetic improvement of pulse crops to develop biotic and abiotic stress-resistant cultivars. However, Thomas *et al*. [3], Girija and Dhanavel [10], Patil *et al*. [13] and Khan *et al*. [74] demonstrated that induced mutagenesis (chemical/ physical) could provide a valuable source of genetic variability for quantitative and qualitative inherent traits in *Phaseolus vulgaris*, *Vigna unguiculata* L., *Glycine max* L., and *Vigna radiata* L. Chemical mutagens such as sodium azide and ethyl methanesulfonate are well known for their mutagenic effects in various legume and non-legume crops [73]. The use of these chemicals already serves as an effective tool for crop improvement, particularly in crops that have narrow genetic bases such as tepary bean.

Genetic variability in a crop species can result from spontaneous natural mutation. However, the rate of spontaneous mutation is low and cannot be exploited for breeding purposes, especially as an alternative to conventional breeding. Classical or traditional breeding that was largely based on the mode of reproduction of the species self-pollinating or cross-pollinating is accompanied by long breeding cycles and a high generation rate of infertility in produced hybrids. Hence artificial mutations are induced with physical and chemical mutagenic agents may serve very beneficial, especially for introgression of genes of interest that are non-existent in nature [50]. Despite the successful application of mutation breeding in a diversity of field crops, there are limited reports on the utilisation of mutation breeding in tepary bean. Therefore, there is merit in the application of these kinds of approaches, for instance, the technique of chemical mutagenesis on tepary bean in order to induce some useful agronomic traits.

As previously explained, chemical mutagen generally produces induced mutations, which lead to nucleotides or base-pair substitution especially GC to AT resulting in amino acid changes, which changes the function of proteins, without abolishing their functions. According to Greene *et al.* [68], EMS mutagenesis can generate 99% of C-to-T changes that result in C/G to T/A substitution. These changes were correlated with the total amount of seed proteins and NRA activity in mung bean and urdbean [73]. However, it was also suggested that such mutational effects could also give rise to many different mutant alleles with varying degrees of trait modifications that are desirable and used for the development of new cultivars, as postulated by Chopra [76] and Kozgar *et al.* [73]. Therefore, induced mutagenesis must be continuously tested in various other cultivated legume grains for their enhancement of the nutritional value, improved agronomic traits and resistance to numerous biotic and environmental stress factors.

CONCLUSION AND FUTURE PROSPECTS

Tepary bean (*Phaseolus acutifolius*) has the greatest potential to play a significant role in the human diet around the world. The crop contains high amounts of protein, carbohydrates, fibre, vitamins, and minerals, associated with preventative effects against various chronic and acute diseases. Although this grain crop is more resistant to high temperatures, sub-zero temperatures, salt stress and drought stress [77], it remains marginally cultivated, and difficult to determine underlying world production, as well as the rates of distributions. However, information and analyses provided in this chapter, particularly based on the growth and yield effects of mutagen EMS have generated additional insights in tepary bean. The findings showed that EMS induced some dominant mutations that were detectable in the M_1 generation of tepary bean seedlings. Mutant generations exhibited significant (P<0.01) variations on seed germination (%), SHT, SDW, SRL and RDW.

There was a negative but highly significant linear relationship between the pod load (as measured by the number of pods per plant) and the number of seeds per pod. The results suggested that in mutation breeding for tepary bean, it is important to consider each of the three factors in order to induce desirable agronomic traits. The study also demonstrated the merits of evaluating mutant genotypes of tepary bean using both the laboratory and field approaches. It would be interesting to expose more diverse germplasm of tepary bean to a broader range of EMS dosages. In addition, it could be interesting to tag the best performing seedlings and retain them for field performance at the matured growth and reproductive stages. Furthermore, these results indicate that mutagen induced characters are highly inheritable, meanwhile, the advanced traits indicate the predominance of additive genes.

Plant mutagenesis can increase the variation in legumes when coupled with high resolution genotypic or phenotypic screening methods, allowing breeders to select for traits of interest [70]. The introduction of new desirable genetic mutations into breeding lines offers a unique opportunity to identify and incorporate novel traits into newly developed cultivars [38, 73]. With the rapidly rising costs and demand for foods, and prolonged drought seasons, accumulation of genetic variability will be beneficially coupled with genome sequencing and high-resolution mutant screening techniques to complement traditional breeding techniques [52]. Furthermore, the tepary bean's relative resistance to biotic and abiotic stress factors, particularly drought, makes it a critical genetic pool for the improvement of other related species. Mwale *et al.* [14] emphasised that targeted selection of agronomic, physiological, and biochemical traits that maximise yield gains using the *Phaseolus* gene pool remains useful to develop stress-tolerant and high-performing genotypes.

LIST OF ABBREVIATIONS

A/T   Adenine-thymine

BNF   Biological nitrogen fixation

EMS   Ethyl methane sulfonate

G/C   Guanine-cytosine

N   Nitrogen

NaM3   Sodium azide

NRA   Nitrogenase reductase activity

RDW   Root dry weight

SDW   Shoot dry weight

SHT   Shoot height

SSA   sub-Saharan Africa

CONSENT FOR PUBLICATION

Not applicable.

CONFLICT OF INTEREST

The author declares no conflict of interest, financial or otherwise.

ACKNOWLEDGEMENTS

Declared none.

REFERENCES

[1]   Nabhan GP, Felger RS. Teparies in south-western North America. Econ Bot 1978; 32: 2-19.

[2]   Mohammed MF, Keutgen N, Tawfik AA, Noga G. Dehydration-avoidance responses of tepary bean lines differing in drought resistance. J Plant Physiol 2002; 159: 31-8.
[http://dx.doi.org/10.1078/0176-1617-00530]

[3]   Thomas CV, Manshardt RM, Waines JG. Teparies as a source of useful traits for improving common beans. Des Plant 1983; 5: 43-8.

[4]   Rao I, Beebee S, Polania J, *et al.* Can tepary bean be a model for improvement of drought resistance in common bean? Afr Crop Sci J 2013; 21: 265-81.

[5]   Salgado MO, Schwartz HF, Brick MA, Pastor-Corrales MA. Resistance to *Fusarium oxysporum* spp. Phaseoli in tepary bean (*Phaseolus acutifolius*). Plant Dis 1994; 78: 357-60.
[http://dx.doi.org/10.1094/PD-78-0357]

[6]   Miklas PN, Rossaa JC, Beaver JS, Telek L, Freytag GF. Field performance of select tepary bean germplasm for the tropics. Crop Sci 1994; 34: 1639-44.
[http://dx.doi.org/10.2135/cropsci1994.0011183X003400060040x]

[7]   Bhardwaj HL, Rangappa M, Hamama AA. Planting date and genotype effects on tepary bean productivity. HortScience 2002; 2: 317-8.
[http://dx.doi.org/10.21273/HORTSCI.37.2.317]

[8]   De Mejia EG, Del Carmen Valadez-Vega M, Reynoso-Camacho R, Loarca-Pina G, Loarca-Pina G. Tannins, trypsin inhibitors and lectin cytotoxicity in tepary (*Phaseolus acutifolius*) and common (*Phaseolus vulgaris*) beans. Plant Foods Hum Nutr 2005; 60(3): 137-45.
[http://dx.doi.org/10.1007/s11130-005-6842-0] [PMID: 16187017]

[9]   Brand JC, Snow BJ, Nabhan GP, Truswell AS. Plasma glucose and insulin responses to traditional Pima Indian meals. Am J Clin Nutr 1990; 51(3): 416-20.
[http://dx.doi.org/10.1093/ajcn/51.3.416] [PMID: 2178389]

[10]   Girija M, Dhanavel D. Mutagenic effectiveness and efficiency of gamma rays, ethylmethanesulfonate and their combined treatments in cowpea. Glob J Mol Sci 2009; 4: 68-75. [*Vigna unguiculata* (L) Walp].

[11]   Gaikwad NB, Kothekar VS. Mutagenic effectiveness and efficiency of ethyl methane sulphonate and sodium azide in lentil. Int J Hum Genet 2004; 64: 73-4.

[12]   Singh R, Kole CR. Effect of mutagenic treatments with EMS on germination and some seedling parameters in mungbean. Crop Res 2005; 30: 236-40.

[13]   Patil A, Taware SP, Raut VM. Induced variation in quantitative traits due to physical (ψ rays), chemical (EMS) and combined mutagen treatments in soybean. Genet 2004; 31: 1-6. [*Glycine max* (L.) Merrill].

[14]   Mwale SE, Shimelis H, Mafonoya P, Mashilo J. Breeding tepary bean (*Phaseolus acutifolius*) for drought adaptation: A review. Plant Breed 2020; 00: 1-13.
       [http://dx.doi.org/10.1111/pbr.12806]

[15]   Woff M. Plant guide for tepary bean (*Phaseolus acutifolius*). USDA-Natural Resource Conservation Service, Tucson Plant Materials Centre Tucson, A2.85705.

[16]   Carroll BJ, McNeil DL, Gresshoff PM. Isolation and properties of soybean [*Glycine max* (L.) Merr.] mutants that nodulate in the presence of high nitrate concentrations. Proc Natl Acad Sci USA 1985; 82(12): 4162-6.
       [http://dx.doi.org/10.1073/pnas.82.12.4162] [PMID: 16593577]

[17]   Shisanya CA. Yield and nitrogen fixation response by drought tolerant tepary bean (*Phaseolus acutifolius* A. Gray var. *acutifolius*) in sole and maize intercrop systems in semiarid Kenya. J Agron 2003; 2: 126-37.
       [http://dx.doi.org/10.3923/ja.2003.126.137]

[18]   Molosiwa OO, Kgokong SB, Makwala B, Gwafila C, Ramokapane MG. Genetic diversity in tepary bean (*Phaseolus acutifolius*) landraces grown in Botswana. J Plant Breed Crop Sci 2014; 6: 194-9.
       [http://dx.doi.org/10.5897/JPBCS2014.0458]

[19]   Gepts P. Beans: Origins and development.Encyclopaedia of global archaeology. New York: Sprinnger 2014; pp. 9-82.
       [http://dx.doi.org/10.1007/978-1-4419-0465-2_2169]

[20]   Cortés AJ, Monserrate FA, Ramírez-Villegas J, Madriñán S, Blair MW. Drought tolerance in wild plant populations: the case of common beans (*Phaseolus vulgaris* L.). PLoS One 2013; 8(5): e62898.
       [http://dx.doi.org/10.1371/journal.pone.0062898] [PMID: 23658783]

[21]   Salas-Lopez F, Gutierrez-Dorado R, Milan-Carrillo J, *et al.* Nutritional and antioxidant potential of a desert underutilized legume-tepary bean (*Phaseolus acutifolius*). Optimization of germination bioprocess. Food Sci Technol (Campinas) 2018; 38 (Suppl. 1): 254-62.
       [http://dx.doi.org/10.1590/fst.25316]

[22]   Maxwell DP, Falk S, Trick CG, Huner N. Growth at low temperature mimics high-light acclimation in *Chlorella vulgaris*. Plant Physiol 1994; 105(2): 535-43.
       [http://dx.doi.org/10.1104/pp.105.2.535] [PMID: 12232221]

[23]   White JW, Montes C. The influence of temperature on seed germination on cultivars of common bean. J Exp Bot 1993; 44: 7595-1800.
       [http://dx.doi.org/10.1093/jxb/44.12.1795]

[24]   Eissenstat DM. Costs and benefits of constructing roots of small diameter. J Plant Nutr 1992; 15: 763-82.
       [http://dx.doi.org/10.1080/01904169209364361]

[25]   Butare L, Rao IM, Lepoivre P, *et al.* New genetic sources of resistance in the genus *Phaseolus* to individual and combined aluminium toxicity and progressive soil drying stresses. Euphytica 2011; 181: 385-404.
       [http://dx.doi.org/10.1007/s10681-011-0468-0]

[26]   Matsui T, Singh BB. Root characteristics in cowpea related to drought tolerance at the seedling stage. Exp Agric 2003; 39: 29-38.
       [http://dx.doi.org/10.1017/S0014479703001108]

[27]   Thangwana A. The response of tepary bean (*Phaseolus acutifolius*) germplasms to induced mutation. Masters Dissertation, University of Venda, South Africa 2016; 18-66.

[28] Mangena P. Phytocystatins and their potential application in the development of drought tolerance plants in soybeans (*Glycine max* L.). Protein Pept Lett 2020; 27(2): 135-44.
[http://dx.doi.org/10.2174/0929866526666191014125453] [PMID: 31612812]

[29] Bhardwaj HL. Preliminary evaluation of tepary bean (*Phaseolus acutifolius* A. Gray) as a forage crop. J Agric Sci 2013; 5: 160-6.

[30] Tull D. Edible and Useful Plants of the Southwest Texas, New Mexico and Arizona. Austin: University of Texas Press 2013; pp. 97-8.

[31] Singh SPS, Munoz CG. Resistance to common bacterial blight among *Phaseolus* species and common bean improvement. Crop Sci 1999; 39: 80-9.
[http://dx.doi.org/10.2135/cropsci1999.0011183X003900010013x]

[32] Conger BV, Skinner LW, Skold LN. Variability for components of yield induced in soybeans by seed treatment with gamma radiation, fission neutrons, and ethylmethanesulfonate. Crop Sci 1976; 2: 281-6.

[33] Miklas PN, Schwartz HF, Salgado MO, Beaver JS. Reaction of select tepary bean to ashy stem blight and *Fusarium* Wilt. HortScience 1998; 33: 136-9.

[34] Goertz Goertz SH, Coons JM. Tolerance of tepary and navy beans to NaCl during germination and emergence. HortScience 1991; 26: 246-9.
[http://dx.doi.org/10.21273/HORTSCI.26.3.246]

[35] Galwey NW, Temple SR, van Schoonhoven A. The resistance of genotypes of two species of *Phaseolus* bean to the leaf hopper *Empoasca kraemeri*. Ann Appl Biol 1985; 107: 146-50.

[36] Shade RE, Pratt RC, Pomeroy MA. Development and mortality of the bean weevil, *Acanthoscelides obtactus* (*Coleoptera: Bruchidae*), on mature seeds of tepary beans, *Phaseolus acutifolious*, and common beans, *Phaseolus vulgaris*. Environ Entomol 1987; 16: 1067-70.
[http://dx.doi.org/10.1093/ee/16.5.1067]

[37] Pratt RC, Singh NK, Shade RE, Murdock LL, Bressan RA. Isolation and partial characterization of a seed lectin from tepary bean that delays bruchid beetle development. Plant Physiol 1990; 93(4): 1453-9.
[http://dx.doi.org/10.1104/pp.93.4.1453] [PMID: 16667639]

[38] Muñoz LC, Blair MW, Duque MC, Tohme J, Roca W. Introgression in common bean x tepary bean interspecific congruity-backcross lines as measured by AFLP markers. Crop Sci 2004; 44: 637-45.
[http://dx.doi.org/10.2135/cropsci2004.6370]

[39] Anderson NO, Ascher PD, Haghighi K. Congruity backcrossing as a means of creating genetic variability in self-pollinated crops: seed morphology of *Phaseolus vulgaris* L. and *Phaseolus acutifolius* A. Gray hybrids. Euphytica 1996; 87: 211-24.
[http://dx.doi.org/10.1007/BF00023748]

[40] Mejía-Jiménez A, Muñoz C, Jacobsen HJ, Roca WM, Singh SP. Interspecific hybridization between common and tepary beans: increased hybrid embryo growth, fertility, and efficiency of hybridization through recurrent and congruity backcrossing. Theor Appl Genet 1994; 88(3-4): 324-31.
[http://dx.doi.org/10.1007/BF00223640] [PMID: 24186014]

[41] Mhlaba ZB, Mashilo J, Shimelis H, Assefa AB, Modi AT. Progress in genetic analysis and breeding of tepary bean (*Phaseolus acutifolius* A. *Gray*): A review. Sci Hortic (Amsterdam) 2018; 237: 112-9.
[http://dx.doi.org/10.1016/j.scienta.2018.04.012]

[42] Porch TG, Beaver JS, Brick MA. Registration of tepary germplasm with multiple stress tolerance, TARS-Tep 22 and TARS-Tep 32. J Plant Regist 2013; 7: 358-64.
[http://dx.doi.org/10.3198/jpr2012.10.0047crg]

[43] Federici CT, Ehdaie B, Waines JD. Domesticated and wild tepary bean: Field performance with and without drought-stress. Agron J 1999; 82: 896-900.

[http://dx.doi.org/10.2134/agronj1990.00021962008200050010x]

[44]   Mohamed MF, Schmitz-Eiberger N, Keutgen N, Noga G. Comparative drought postponing and tolerance potentials of two tepary bean lines in relation to seed yield. Afr Crop Sci J 2005; 13: 49-60.

[45]   Beebe SE, Rao IM, Blair MW, Acosta-Gallegos JA. Phenotyping common beans for adaptation to drought. Front Physiol 2013; 4(35): 35.
[http://dx.doi.org/10.3389/fphys.2013.00035] [PMID: 23507928]

[46]   Nabhan GP, Felger RS. Teparies in southwestern North America. Econ Bot 1978; 32: 2-19.
[http://dx.doi.org/10.1007/BF02906725]

[47]   Anderson JM, Ingram JSI. Tropical soil biological and fertility: A handbook of methods. 2nd ed. Wallingford, UK: CAB International 1993; p. 232.

[48]   Biswas B, Gresshoff PM. The role of symbiotic nitrogen fixation in sustainable production of biofuels. Int J Mol Sci 2014; 15(5): 7380-97.
[http://dx.doi.org/10.3390/ijms15057380] [PMID: 24786096]

[49]   Sears ER. Use of radiation to transfer alien segments to wheat. Crop Sci 1993; 33: 897-901.
[http://dx.doi.org/10.2135/cropsci1993.0011183X003300050004x]

[50]   Mba C, Afza R, Bado S, Jain SM. Induced mutagenesis in plants using physical and chemical agents.Plant cell culture: Essential methods. UK: Wiley 2010; p. 136.
[http://dx.doi.org/10.1002/9780470686522.ch7]

[51]   Predieri S. Mutation induction and tissue culture in improving fruits. Plant Cell Tissue Organ Cult 2001; 64: 185-21.
[http://dx.doi.org/10.1023/A:1010623203554]

[52]   Mangena P. *In vivo* and *in vitro* application of colchicine on germination and shoot proliferation in soybean. Asian J Crop Sci 2020; 12: 34-42. [*Glycine max* (L.) Merr.].
[http://dx.doi.org/10.3923/ajcs.2020.34.42]

[53]   Ndou VN, Shimelis H, Odindo A, Modi AT. Response of selected wheat genotypes to ethyl methane sulfonate concentration, treatment temperature and duration. Sci Res Essays 2013; 8: 198-6.

[54]   Horn L, Shimelis H. Radio-sensitivity of selected cowpea (*Vigna unguiculata*) genotypes to varying gamma irradiation doses. Sci Res Essays 2013; 8(40): 1991-7.

[55]   Lukanda TL, Funny-Biola C, Tshiyoyi-Mpunga A, *et al.* Radio-sensitivity of some groundnut (*Arachis hypogaea* L.) genotypes to gamma irradiation: indices for use as improvement. Brit J Biotech 2012; 3: 169-78.
[http://dx.doi.org/10.9734/BBJ/2012/1459]

[56]   Gnanamurthy S, Dhanavel D. Effect of EMS on induced morphological mutants and chromosomal variation in Cowpea. Int Lett Nat Sci 2014; 17: 33-43. [*Vigna unguiculata* (L.) Walp].

[57]   Tah PR. Studies on gamma ray induced mutations in mung bean. Asian J Plant Sci 2006; 5: 61-70. [*Vigna radiata* (L.) Wilczek].

[58]   Khan S, Goyal S. Improvement of mung bean varieties through induced mutations. Afr J Plant Sci 2009; 3: 174-80.

[59]   Essel E, Asante IK, Laing F. Effect of colchicine treatment on seed germination, plant growth and yield traits of cowpea (*Vigna unguiculata* (L.) Walp). Canad J Pure. Appl Sci (Basel) 2015; 9(3): 3573-6.

[60]   Verhoeven T, Fahy B, Leggett M, Moates G, Denyer K. Isolation and characterisation of novel starch mutants of oats. J Cereal Sci 2004; 40(1): 69-79.
[http://dx.doi.org/10.1016/j.jcs.2004.04.004]

[61]   Jamil M, Khan UQ. Study of genetic variation in yield components of wheat cultivar Bukhtwar-92 as induced by gamma radiation. Asian J Plant Sci 2002; 5: 579-80.

[http://dx.doi.org/10.3923/ajps.2002.579.580]

[62]   Muhammad A, Rasul E, Muhammad S. Mutagenic response of *Macrosperma* lentils to gamma rays. Pak J Biol Sci 2000; 3: 1605-8.
       [http://dx.doi.org/10.3923/pjbs.2000.1605.1608]

[63]   Çelik O, Atak C, Suludere Z. Response of soybean plants to gamma radiation: Biochemical analyses and expression patterns of trichome development. Plant Omics J 2014; 7: 382-91.

[64]   Asghari R, Razavi A, Bakhtiari S, Soleymanifard S. Germination of X-ray treated wheat seeds in saline condition. Int J Agric Crop Sci 2013; 6: 1153-63.

[65]   Till BJ, Reynolds SH, Weil C, *et al.* Discovery of induced point mutations in maize genes by TILLING. BMC Plant Biol 2004; 4: 12.
       [http://dx.doi.org/10.1186/1471-2229-4-12] [PMID: 15282033]

[66]   Okagaki RJ, Neuffer MG, Wessler SR. A deletion common to two independently derived waxy mutations of maize. Genetics 1991; 128(2): 425-31.
       [PMID: 2071021]

[67]   Henry IM, Nagalakshmi U, Lieberman MC, *et al.* Efficient genome wide detection and cataloguing of EMS-induced mutations using exome capture and next generation sequencing. Plant Cell 2014; 26(4): 1382-97.
       [http://dx.doi.org/10.1105/tpc.113.121590] [PMID: 24728647]

[68]   Greene EA, Codomo CA, Taylor NE, *et al.* Spectrum of chemically induced mutations from a large-scale reverse-genetic screen in *Arabidopsis.* Genetics 2003; 164(2): 731-40.
       [PMID: 12807792]

[69]   Hartl Dl, Jones EW. Molecular mechanisms of mutation and DNA repair: Genetics Analysis of Genes and Genomes. 5th ed. Boston, USA: Jones and Bartlett Publishers 2001; pp. 264-309.

[70]   Sikora P, Chawade A, Larsson M, Olsson J, Olsson O. Mutagenesis as a tool in plant genetics, functional genomics, and breeding. Int J Plant Genomics 2011; 1-14. ID 314829

[71]   Sangsiri C, Sorajjapinun W, Srinives P. Gamma radiation induced mutations in mungbean. Sci Asia 2005; 31: 251-5.
       [http://dx.doi.org/10.2306/scienceasia1513-1874.2005.31.251]

[72]   Anbarasan K, Sivalingam D, Rajendran R, Anbazhagan M, Chidambaram AA. Studies on the mutagenic effect of EMS on seed germination and seedling characters of Sesame (*Sesamum indicum* L.) Var. T MV3. Int J Res Biol Sci 2013; 3(1): 68-70.

[73]   Kozgar MI, Goyal S, Khan S. EMS induced mutational variability in *Vigna radiata* and *Vigna mungo.* J Bot (Faisalabad) 2011; 6: 31-7.
       [http://dx.doi.org/10.3923/rjb.2011.31.37]

[74]   Khan S, Wani MR, Parveen K. Induced genetic variability for quantitative traits in *Vigna radiata* (L.) Wilczek. Pat J Bot 2004; 36(4): 845-50.

[75]   Wani MR, Khan S, Parveen K. Induced variation for quantitative traits in mungbean. Ind J Applied Pure Biol 2005; 20: 55-8.

[76]   Chopra V. Mutagenesis: Investigating the process and processing the outcome for crop improvement. Curr Sci 2005; 89: 353-9.

[77]   Souter JR, Gurusamy V, Porch TG, Bett KE. Successful introgression of abiotic stress tolerance from wild tepary bean to common bean. Crop Sci 2017; 57: 1160-71.
       [http://dx.doi.org/10.2135/cropsci2016.10.0851]

CHAPTER 6

Effect of Drought Stress on Soybean Nodule Proteomes and Expression of Cysteine Proteases

Phumzile Mkhize[*]

Department of Biochemistry, Microbiology and Biotechnology, School of Molecular and Life Sciences, Faculty of Science and Agriculture, University of Limpopo, Sovenga, 0727, South Africa

Abstract: Soybean is a valuable crop cultivated as an excellent source of proteins, dietary fibre, and a variety of micronutrients. *Rhizobia* reside as symbiosomes in the infected cells of soybean nodules to fix atmospheric nitrogen. Such association ensures optimal yield and may be beneficial in quenching the use of fertilizers. However, premature nodule senescence remains one of the major problems affecting soybean growth and yield. A clear understanding of the molecular level on the resultant effects of factors that promote early nodule senescence is important. Cysteine proteases are the proteases directly involved in the commencement of early tissue senescence. Nevertheless, the studies performed on the involvement of these proteases in nodule senescence have not reached a consensus. Besides, the specific family and isoforms of cysteine proteases expressed under normal well-watered, waterlogging and water deficit conditions during nodule senescence are not clearly understood. Consequently, there is a need for conducting intensive molecular studies on the involvement of cysteine proteases in nodule senescence. Changes in the whole proteome of a healthy and a senescing nodule need to be understood to make conclusive findings on the molecular events that occur during senescence. The use of proteomic approaches to investigate the level and characteristics of the proteins in general, by preparing a protein map and its application to functional analysis in soybean nodule senescence is important in delaying premature senescence.

Keywords: Cysteine proteases, Cystatins, Drought stress, Nodule senescence, Proteomics, *Rhizobia*, Soybean.

INTRODUCTION

Rhizobia reside as symbiosomes in the infected cells of the legume nodules where they fix atmospheric nitrogen. During this relationship, the soybean plant provides the energy required by the bacterium in the form of sucrose that is produced during photosynthesis.

---

[*] **Corresponding author Phumzile Mkhize**: Department of Biochemistry, Microbiology and Biotechnology, School of Molecular and Life Sciences, Faculty of Science and Agriculture, University of Limpopo, South Africa; Tel: +2715 268 3017; E-mails: Phunzile.Mkhize@ul.ac.za & maphumup88@gmail.com

Phetole Mangena (Ed.)

In turn, the bacterium produces ureides, which get converted into amino acids for subsequent use by the plant in protein synthesis [1]. This symbiotic relationship is well regulated and may last for some time until the entry of the nodule into the senescence stage. Abiotic stress, such as drought, is one of the main factors that led to the early senescence of soybean nodules resulting in low grain yields [2]. In light of the recent changes observed in the climate, it is imperative that abiotic stress and factors associated with it should be given more focus to ensure maximal grain yield from soybean and other legume plants [3].

Premature senescence in soybean nodules can negatively affect the availability of nitrogen required for plant growth and quality and quantity of yield. A clear understanding, at the molecular level, of the resultant effects of factors that promote early nodule senescence is, therefore, significant. The research focus should be driven towards investigating delayed nodule senescence since it has been proven that nodulation results in improved crop productivity [4]. Cysteine proteases are some of the proteases that are directly involved in initiating premature nodule senescence [2, 5 - 7]. However, some researchers still argue their involvement, consequently, creating a need to conduct intensive molecular research on their involvement in nodule senescence. Changes in the whole proteome of both healthy and senescing nodules need to be evaluated to make conclusive findings.

An understanding of the proteins that are expressed during nodule senescence would thus lead to the elucidation of the mechanisms that are involved during senescence and offer an opportunity to regulate stress-induced nodule senescence [8]. Besides, the specific family and isoforms of cysteine proteases expressed under well-watered conditions, waterlogging, and water deficit conditions in nodule senescence are also not clearly understood. Therefore, the objective of this review is to highlight recent progress on the topic and to provide insights into the nature of proteases that are responsible for the degradation of proteins during drought stress-induced senescence in nodules. Other proteins that are associated with legume senescence and their possible role are also discussed.

ECONOMIC IMPORTANCE OF SOYBEAN

Soybean has been investigated extensively for its nutritional attributes in the legume group [9, 10]. It serves as an excellent source of phytochemicals such as the isoflavones, which are a subclass of flavonoids. The anticancer property of the 2 primary isoflavones found in soybean has received the most attention among the phytochemicals found within this plant [11, 12]. Soybean is processed into a variety of foods such as soya milk, sausage, cheese, and yoghurt. It plays an important role in traditional diets throughout the world, especially in developing countries [13]. Soybean serves as an excellent source of proteins, dietary fibre, a variety of micronutrients, phytochemicals and it is also low in fats. A total of 100 g of soybean seed powder has been estimated to contain 173 Kcal energy, 17 g proteins, 1 g salt, 10 g carbohydrate, and 6 g fibre. The amount of nutritional value indicates the importance of using soybeans as a good source of proteins [12]. This crop requires 120 to 130 days from planting to maturity. The duration for maturity only applies where the plant is well adapted to the environment and in soils where a proper balance of macro-micro nutrients is maintained.

The use of fertilisers for the supply of mineral nutrients is widely accepted. However, these may be beyond the reach of most small scale farmers in developing countries. The use of biological nitrogen fixation (BNF) provides efficient levels of nitrogen (N_2), especially as an alternative to reduce the application of nitrogenous fertilizers. Research has estimated an annual N_2 contribution between 25 to 30% from nitrogen-fixing organisms in areas such as Sub-Sahara Africa and eastern Australia [14, 15]. However, with the improvement in the proper management of the BNF, it is further estimated that in the future these can reduce the money spent on fertilizers by an estimate of 80 to 90 million Rands [16]. The use of fertilizers does not only cause monetary losses but also poses a threat to the environment, instigating water, and air pollution, particularly in increasing the amount of nitrates in drinking water [10]. Crews and the group [9] concluded that obtaining fixed N_2 from legumes is potentially more sustainable and environmentally friendly than sourcing it from industries in the form of fertilisers.

PROTEOMIC APPROACH FOR SOYBEAN CROP IMPROVEMENT

Plants growing under uncontrolled natural habitats are frequently subjected to different stress factors and could require metabolic adjustments or specific stress-induced proteins for adaptation. Molecular biosynthesis of enzymes used by plants as osmoprotectants, enzymes involved in antioxidant activity, and heat shock proteins (chaperones) are some of the proteins required for protection in plants [17]. Several attempts have been made to create protein maps using two-dimensional sodium dodecyl sulfate-polyacrylamide gel electrophoresis (2-D SDS-PAGE) and matrix-assisted laser desorption/ionisation time-of-flight mass spectrometry (MALDI-TOF MS) (Table **1**). These protein maps are significant for providing more insights into molecular mechanisms, protein identity, and expression levels, as well as possible roles in soybean development and senescence under stress conditions. At another level, establishing these maps will offer an opportunity to regulate the expression of these proteins using genetic engineering approaches.

Soybean plants are also susceptible to environmental stress, particularly drought, that causes dehydration characterised by profound changes in the water availability for fundamental biochemical processes, membrane structure and subcellular organelles. For instance, the endoplasmic reticulum (ER) is the central organelle that regulates stress responses in plant cells. It plays a critical role in integrating signals generated by abiotic stress in plants by processing massive reprograming of transcription and translation of stress response regulators [18, 19]. The demand for high protein synthesis may affect the quality of protein synthesis leading to the generation of unfolded and/or misfolded proteins in the ER and cytosol, which is the most detrimental effect of stress in plants. Therefore, proteins that ensure homeostasis in the ER are essential in enhancing stress tolerance, like the ER-resident molecular chaperone Bip (binding protein).

Valente *et al.* [20] reported that transgenic soybean plants overexpressing Bip showed increased tolerance to drought stress by retaining water potential and exhibiting less leaf wilting compared to the wild type. Identification and analysis of more proteins whose expression is enhanced by abiotic stress in general, especially under drought-prone environments are vital in combatting the effect of stress in soybean. Mohammadi *et al.* [30] also reported the reduction in protein abundance due to drought as a characteristic of aging during legume symbiosis. Additionally, van de Velde *et al.* [31] indicated that mass degradation of protein is a characteristic of nodule senescence due to the loss of hemoprotein and leghaemoglobin.

Table 1. The use of 2-D SDS-PAGE and MALDI-TOF MS proteomic approach for soybean development and functional analysis through protein expression under different stress conditions.

Stress Condition	Target Tissue	Identified Proteins	Reference
Salinity, Flooding	Hypocotyl, Roots	Upregulation of proteins associated with secondary metabolites, enzymes catalysing glycolysis and fermentation pathway.	Alam *et al.* [21]
None	Peroxisomes	Identified enzymes for fatty acid β-oxidation, glyoxylate cycle, photorespiratory glycolate metabolism, stress response and metabolite transport	Arai *et al.* [22]
None	Nodule Mitochondria	Identified phosphoserine aminotransferase, flavanone 3-hydroxylase, coproporphyrinogen III oxidase and ribonucleoprotein	Hoa *et al.* [23]
None	Nodule Peribacteriod Membrane	Chaperones (HSP60, Bip [HSP70] and PDI), serine and cysteine protease were obtained as proteins associated with the peribacteriod	Panter *et al.* [24]
UV-Light	Leaves	Proteins involved in metabolism, energy, secondary metabolites, disease and defence were upregulated	Xu *et al.* [25]

(Table 1) cont.....

Stress Condition	Target Tissue	Identified Proteins	Reference
Aluminium Toxicity	Roots	Up regulation of heat shock protein, glutathione S-transferase, chalcone-related synthetase, GTP-binding protein, ABC transporter ATP-binding protein	Zhen *et al.* [26]
Salt	Roots	Up regulation of β-conglycinin and elicitor peptide three precursor. Down regulation of protease inhibitor	Aghaei *et al.* [27]
Flooding	Seedlings	Cytosolic ascorbate peroxidase 2	Shi *et al.* [28]
Cadmium Stress	Cell suspension culture	Up regulation of superoxide dismutase, histone H2B, chalcone synthase and glutathione transferase	Sobkowiak and Deckert [29]

These observations are in line with the results displayed in Fig. (**1A**), where the total soluble protein concentration was reduced, with most of the prominent proteins expressed over the entire duration. However, the use of 2-D SDS-PAGE showed a reduction in the number of protein spots and intensity between water-stressed and non-stressed plant over time (Fig. **1B - E**). A total of 155 protein spots were subjected to analysis using MALDI-TOF MS, in which a total of 6 protein spots were positively identified as indicated by number 1-6 in Fig. (**1B, 1C**). In line with the previous reports, the study revealed that nucleosome assembly protein 1 involved in DNA replication and cell proliferation was down-regulated on nodules subjected to drought stress.

Moreover, heat shock factor was upregulated (Table **2**), a pattern that was also reported by Mohammadi *et al.* [30] in drought-stressed soybean plants. Such proteins are produced by cells in response to exposure to environmental stress. It would be interesting to identify the protein spots in Fig. (**1D and 1E**) that appeared to contain molecular weights and isoelectric points as those of cysteine proteases. Understanding the mechanism triggered by these proteins and the identification of more drought-related proteins are vital for ensuring that methods of preventing the impact of drought stress are developed. Such insights will guide the selection and breeding of legume crops that exhibit specific stress-resistant proteins, thus ensuring maximal yields, accompanied by optimal pod filling and subsequent provision of good quality proteins and oils.

Cysteine Proteases

Proteases are classified according to their catalytic site and categorized into four classes, *i.e.,* cysteine proteases, serine proteases, aspartic proteases, and metallo proteases [33]. In the case of cysteine proteases, the active site contains a Cys-His-Asn triad. This act as a proton donor, enhancing the nucleophilic attack of the peptides during peptide bond hydrolysis. Cysteine proteases are compart mentalized to prevent uncontrolled proteolysis, often taking place during exposure of plants to biotic and abiotic stress. They play multiple roles including remobilisation of protein from dead tissue to use on newly developing ones, killing of invading microorganisms, cellular homeostasis and modulate nodule senescence [34]. These enzymes are also involved in almost all growth stages such as germination, senescence, and programmed cell death. In soybean, cysteine proteases are expressed during the development of determinate nodules located in root nodule regions where BNF starts to decline [35].

Table 2. Identification of protein spots subjected through a BLAST search on the NCBI with molecular functions.

Protein Spot No.	Score	Protein Name	Protein Function/s
1	38	Heat shock factor 30	DNA-binding protein that binds heat shock promoter elements and activates transcription
2	26	Defensin-like protein 266 OS	Killing of cells from other organisms
3	28	Nucleosomes assemble protein 1	Involved in DNA replication and regulation of cell proliferation
4 and 5	23	NAD (P)H dehydrogenase	Prevents the formation of radical species
6	33	50S ribosomal protein L14	Large ribosomal subunit. Involved in translation

Papain-like cysteine proteases are some of the most widely studied cysteine proteases in plants due to their involvement in tissue senescence and programmed cell death. These proteases are synthesised in the endoplasmic reticulum as zymogens (inactive form) and transferred to different organelles only in the presence of environmental stimuli. Upon reaching the target organelles, they are processed into active forms by vacuolar processing enzymes (VPE). VPEs also referred to as legumains or asparaginyl endopeptidase (AEP) are located in the cell walls where they recognise the aspartic acid as part of the target sequence. Therefore, targeting papain-like cysteine proteases on multiple stages during nodule development would generate useful molecular data required for trait improvement of legume crops [37, 38]. Their activation by auto-processing triggered by the change in pH (to acidic pH) involves the removal of the pro-domain that blocks the active site.

VPEs recognise and cleave the peptide bonds on the C-terminal side of the aspartic acid residue [33, 39]. Similar to other cysteine proteases VPEs have Cys and His residues on the active site but they are not sensitive to inhibition by E-64, a common cysteine protease inhibitor. The importance of the activation of papain-like cysteine proteases is further supported by the observations made by Cilliers *et al.* [40] that legumain-like cysteine protease mutant plants showed less papain proteolytic activity (Table **3**). Therefore, targeting VPEs might be the tool for controlling premature senescence caused by papain-like cysteine proteases in nodules [41].

Table 3. Recently published studies on the involvement of cysteine proteases in soybean nodule senescence.

Study	Study Outcome/s	Reference/s
Role of cysteine proteases and protease inhibitor in programmed cell death	• Exogenous application of nitrogen at preliminary stages of senescence improved yields and seed protein content. Concluded that delaying senescence would have the same effect. • Silencing cysteine-protease gene delayed nodule senescence, extended nitrogen fixing period and produced larger nodules in *Astragalas Sinicus* • Ectopic expression of cystatin resulted in drastic reduction in senescence whereas application of inhibitors to serine protease proved ineffective	Sassi-Audi *et al.* [4], Solomon *et al.* [36], Li *et al.* [42]
Roles for a cysteine protease and hydrogen peroxide in nodule senescence	• up-regulation of cysteine proteases that correlated with the onset of nodule senescence. • Papain-like cysteine proteases as the proteases behind nodule senescence	Alesandrini *et al.* [5]
Analysis of cDNA encoding cysteine proteases	• non-uniform expression of genes that codes for cysteine proteases is the main problem in nodule senescence studies	Drake *et al.* [43]
Analysis of cysteine protease cysteine-proteases inhibitor system in nodule	• high expression of cysteine proteases in young nodules than in senescing one. • Legumain-like cysteine proteases as the proteases behind nodule senescence • Obtained 8 isoforms of papain-like cysteine proteases in young nodules and only 2 in senescing one.	Vorster *et al.* [44]
Identification and analysis of changes of drought-induced cysteine protease transcriptome	• Eight and three papain-like and legumain-like cysteine proteases were induced by drought stress, respectively. • Papain-like proteases were highly induced during drought-triggered senescence when compared to the one occurring naturally • Silencing of legumain like cysteine protease reduced the protease activity of papain like cysteine protease and resulted in her biomass and increased protean levels.	Cilliers *et al.* [40], Mangena [45]
Characterization of a root nodule-specific cysteine proteinase cDNA from soybean	• The isolated cDNA clone (GmCysP1) was a soybean nodule-specific belonging to the legumain family was classified as the main protease for nodule senescence.	Oh *et al.* [46]

SENESCENCE IN NODULATED PLANTS

A number of researchers have investigated the cascade of events that lead to early nodule senescence under normal and environmental stress conditions [2, 31, 47]. Microscopic analysis indicated that nodule senescence is marked by the degradation of the peribacteriod membrane, followed by the release of the cellular components. According to Puppo *et al.* [2] and van de Velde *et al.* [31], senescing cells gradually become reabsorbed, while microbial cells show signs of degradation immediately after the death of plant cells. Consequently, the loss in symbiotic relationship prohibits the fixation of nitrogen by the roots, resulting in reduced plant growth, yield and plant death, which is a major problem in the agriculture of legume crops [31]. Genetic studies on soybean nodules indicated that almost all protease families are involved in the senescence process. Most studies have reported the up-regulation of cysteine proteases correlated with the onset of nodule senescence, and these become more pronounced as the senescence process progresses [5, 48, 49].

There are indications for more research to be conducted to ensure that only a specific class of proteases are targeted to prevent premature nodule senescence. In our studies, the presence of cysteine proteases was confirmed through purification using a column packed with thiopropyl sepharose 6B linked to E-64. A pure protein band was observed at an expected size of approximately 25 to 30 kDa (Fig **2A**). The same protein band was obtained in the crude samples separated using one dimensional SDS-PAGE. Cysteine protease expression also appears as a result of senescence in nodules [44]. Lidgett *et al.* [50] reported cysteine proteases as active proteases that are differentially expressed when rapid changes in cell metabolism are required. Using transcriptome analysis, multiple genes that code for various proteases were obtained in both leaves and nodules [44]. These proteases were directly involved in macronutrient degradation and remobilizing. Amongst the expressed proteases, it is clear that cysteine proteases are the most prominent, especially in the initial stages of senescence, further indicating the need for more research on the biochemical processes that activate cysteine proteases.

Detection by streptavidin blotting has not yet been widely reported. In this study, the protein band observed at approximately 25-30 KDa was evidence of the presence of active cysteine proteases (Fig. **2B**, lane **2** and **4**). Addition of E-64 blocked binding DCG-04 as it acts as a competitive inhibitor for cysteine proteases and competed with affinity tag as its supported by absence of signal in Fig. (**2B**, lane **3** and **5**). The insensitivity of legumain like cysteine proteases to E-64 could perhaps indicate that the cysteine protease isolated in water-stressed plants are papain in nature and not legumain (Fig. **2B**). The use of similar techniques will permit the assessment of active cysteine proteases for profiling. Moreover, it would be interesting to check and quantify if non-stressed plants express the same proteases. Contrary to the previous research Vorster *et al.* [44] reported higher expression of cysteine proteases in young nodules than in senescing ones, and they also reported the legumain-like cysteine proteases as the proteases behind nodule senescence.

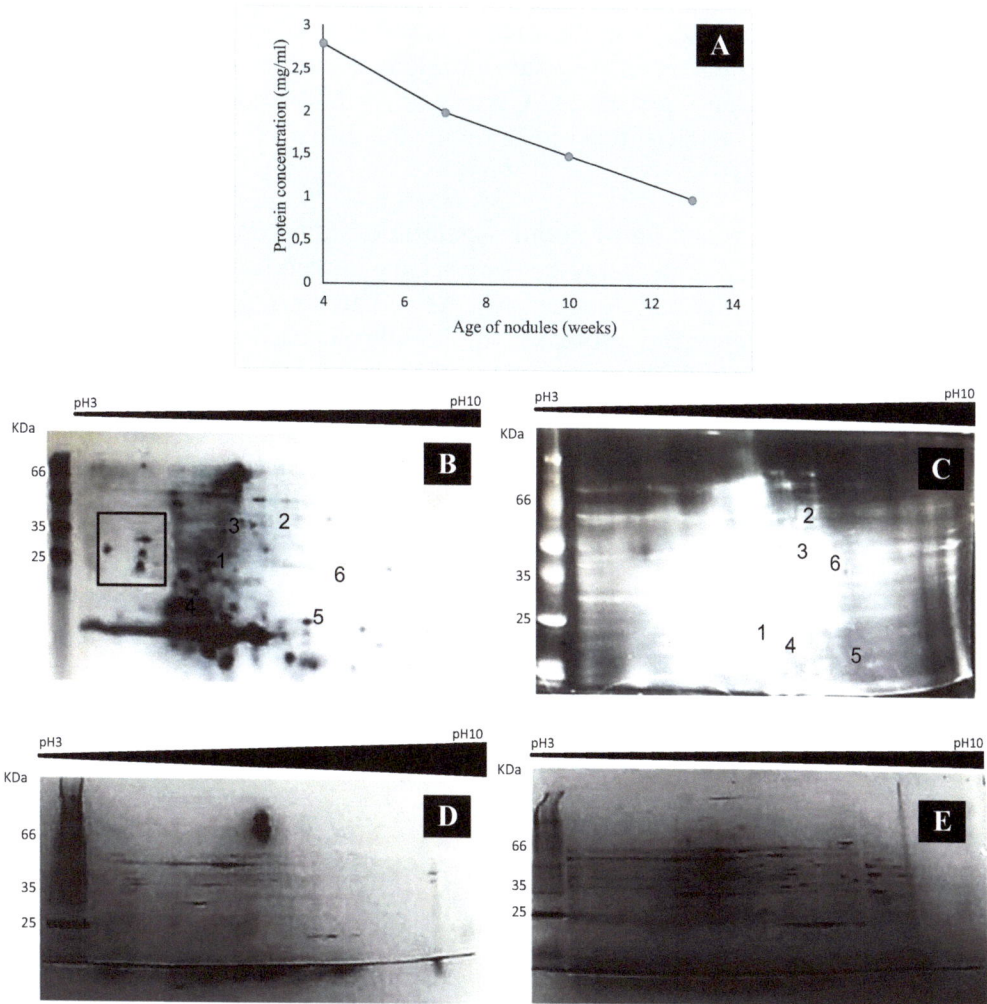

Fig. (1). (A) - Total soluble protein concentration in water-stressed soybeans over time estimated through Bicinchoninic acid assay (BCA) using nodule extract according to Smith *et al.* [32]. Bovine serum albumin (BSA) at different concentrations (0.5 – 5 mg/ml) was used as the standard protein. **(B and D)** - 2D analysis of proteins expressed in nodules during induced drought stress. **(C and E)** - Nodule protein profiles obtained in well-watered soybean plants.

In addition, Vorster *et al.* [44] reported more cysteine protease isoforms in younger nodules than the aged ones. This observation is in line with our findings, elucidating the more intense protein spots observed in Fig. (**1B**) when compared to D. According to these findings, the molecular weight and isoelectric point of these protein spots correlate closely with those of cysteine proteases. Thus, the ultimate goal of our studies would be to identify these protein spots and confirm their function using MALDI-TOF MS and through the use of immuno-detection techniques. Drake *et al.* [43] also reported a non-uniform expression of genes that codes for cysteine proteases as the main problem in studies concerning the link between senescence and proteases. Conceivably this could also explain the difference in protein band intensities previously observed in Fig. (**1B**) when compared to Fig. (**1D**).

Methods such as the silencing of nodule-specific cysteine proteinase genes have proved to be successful in delaying senescence, extending nitrogen fixation period, and enlarging the nodule size [42]. As research continues to indicate the role played by the cysteine proteases in nodule senescence (Table **3**), it is therefore important to elucidate the impact of these proteases by either ensuring an up-regulation of inhibitors such as cystatins or the use of genetically selected crops.

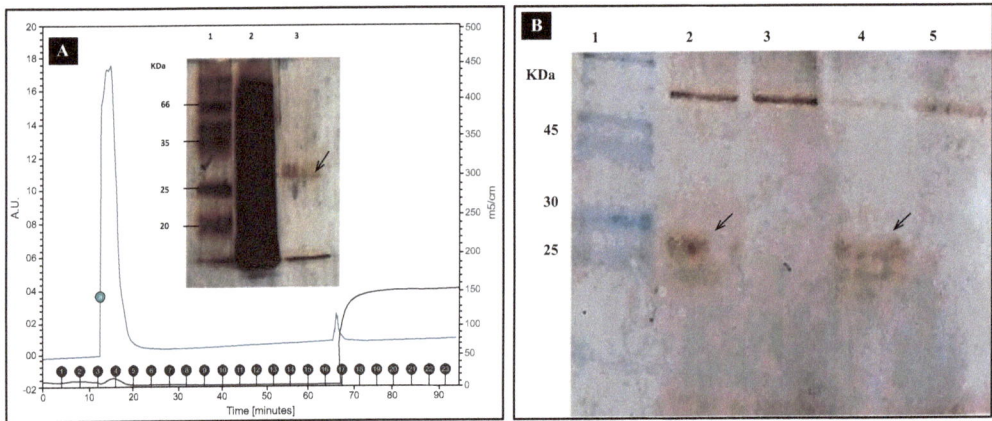

Fig. (2). Cysteine purification using thiopropyl sepharose 6B linked to E-64 (**A**). Probing cysteine proteases in crude extract using DCG-04 (**B**).

Cystatin Expression as a Tool to Understand Cysteine Protease Involvement in Senescence

Protease inhibitors are classified according to the type of protease they inhibit and by the mechanism of action. Inhibitors of cysteine proteases are the cystatin superfamily including the stefins, the cystatins, the kininogens and the phytocystatins. These inhibitors are involved in the regulation of endogenous protein turnover during growth and developmental processes. Also, they are involved in the regulation of senescence, programmed cell death and accumulation and mobilisation of storage proteins [51 - 53]. The tertiary structure of oryzacystatin, the first well characterized cystatin in plants, consists of a five-stranded antiparallel β-sheet wrapped around central α-helix. The structure is well-engineered such that it promotes a perfect fit to the active site of the cysteine protease [54]. Cystatins are natural and specific inhibitors of cysteine proteases of the papain C1 A family.

Cystatins block the C1 proteolytic activity by forming a tight and reversible equimolar complex interaction with cysteine proteases, acting as pseudo-substrate to penetrate the active site of the target enzyme. As such these inhibitors are of particular interest because they form complexes with the target enzyme even after their catalytic site has been inactivated, a phenomenon not reported in other inhibitors [51, 55]. The inhibitory element by cystatins is achieved through the formation of a tripartite wedge into the active site of the cysteine protease. The first specific and physical interaction between the enzyme and the inhibitor is through a hairpin loop of a conserved region bearing Gln-Xaa-Val-Xaa-Gly motif. The second interaction is through the Pro (Leu) – Try motif. Lastly, a conserved Gly residue ensures strong attachment and inhibitory activity [45, 56].

The role of cystatin in nodule senescence has been largely discussed with the aim of using them as a tool for stopping premature nodule senescence in different crops, particularly in soybean. Cystatins were actively transcribed during the nodule development stages; however, the presence of non-transcribed cystatins indicated that these might act as reservoirs for response in certain environments [57]. Similarly, an upregulation of cystatin mRNA transcripts in soybean leaves was observed, particularly N2 and R1 cystatin when the plant was subjected to wounding and methyl jasmonate. Ectopic phytocystatin expression increases nodule numbers and enhanced tolerance to nitrogen limitation, and showed even greater levels of expression when the nodules were entering into the senescing stage [58, 59]. Furthermore, the protective role of cystatin in plants was confirmed when ectopically expressed cystatins were used in transgenic plants showing tolerance to salt, drought, oxidative stress, and chilling when compared to the non-transformed plants.

When the genes encoding for these cystatins were expressed and purified from bacterial cells they inhibited the activity of the papain-like cysteine proteases, which emulsifies the involvement of papain-like cysteine proteases during the senescence stages [57, 60, 61]. Consequently, some of the cystatin can be referred to as inducible cystatins as they are stored during plant development, and their transcription is activated by the signal from the environmental stress and mechanical wounding. Furthermore, the high expression of cystatin upon the commencement of the senescence process indicates that they can be good candidate for reducing the activity of cysteine proteases and proves that controlled expression of inhibitor genes for cysteine proteases can delay nodule senescence in soybeans [36].

RECOMMENDATIONS FOR DELAYING SENESCENCE THROUGH TARGETING CYSTEINE PROTEASES

The grouping of cystatin members into their specific roles in nodule symbiotic signal transduction, modulation, nodule development, and senescence need to be conducted. This will ensure that specifically targeted cystatins are expressed to delay nodule senescence [62]. Such advances will enable the efficient development of genetically modified plants showing resistance to biotic and abiotic stress factors. Soybean plants expressing multiple engineered cystatins for controlled modulation of a number of specific key proteolytic processes that do not interfere with non-targeted ones could be beneficial. According to van Wyk *et al.* [57], this is achievable through amino acid substitution in cystatin motif regions that direct specific binding. Advanced knowledge on the nature of protease isoforms that are responsible for the degradation of essential proteins during drought stress-induced senescence in nodules and leaves also remain critically significant.

A thorough understanding of these protease isoforms will assist in future research where legume plants may be selected on the basis of expression levels of specific proteases that facilitate senescence. Moreover, the knowledge of these isoforms will simplify and accelerate the use of cystatins showing high affinity serving as active isoforms. Confirmatory studies on the class of cysteine proteases directly involved in nodule senescence need to be conducted as contradictory information has been reported on the involvement of both legumain-like and papain-like cysteine proteases [33, 44, 46]. This will allow for the specific plant selection to avoid premature senescence. The effect of generating mutant plants needs to be investigated as recent reports by Cilliers *et al.* [40] indicated that silencing legumain-like gene reduces the protease activity of papain-like protease in drought stressed plants during senescence.

CONCLUSION

Soybean serves as an important food and fodder crop in the agricultural sector. The exploitation of proteomic analysis with the aim of developing a protein map is a starting point to fully understand different molecular biosynthesis, especially under drought stress conditions. Moreover, continued research towards developing a protein map should permit rapid comparison between soybean lines and facilitate the selection of drought-tolerant cultivars expressing proteins of interest. The research focussed on the importance of cysteine proteases and cystatins in promoting or reducing nodule senescence will form the basis for effective strategies in engineering stress-resistant soybeans.

LIST OF ABBREVIATIONS

BCA	Bicinchoninic acid assay
Bip	Binding proteins
BNF	Biological nitrogen fixation
Cys	Cysteine
DNA	Deoxyribonucleic acid
1-D-SDS-PAGE	One-dimensional sodium dodecyl sulfate-polyacrylamide gel electrophoresis
2-D-SDS-PAGE	Two-dimensional sodium dodecyl sulfate-polyacrylamide gel electrophoresis
ER	Endoplasmic reticulum
His	Histidine
MALDI-TOF MS	Matrix-assisted laser desorption/ionisation time-of-flight mass spectrometry
N_2	Nitrogen
VPE	Vacuolar processing enzymes

CONSENT FOR PUBLICATION

Not applicable.

CONFLICT OF INTEREST

The author declares no conflict of interest, financial or otherwise.

ACKNOWLEDGEMENTS

Declared none.

REFERENCES

[1] Wang Q, Liu J, Zhu H. Genetic and molecular mechanisms underlying the symbiotic specificity in legume *Rhizobium* interactions. Front Plant Sci 2018; 9: 313.
[http://dx.doi.org/10.3389/fpls.2018.00313] [PMID: 29593768]

[2] Puppo A, Groten K, Bastian F, *et al.* Legume nodule senescence: roles for redox and hormone signalling in the orchestration of the natural aging process. New Phytol 2005; 165(3): 683-701.
[http://dx.doi.org/10.1111/j.1469-8137.2004.01285.x] [PMID: 15720680]

[3] Ahuja I, de Vos RCH, Bones AM, Hall RD. Plant molecular stress responses face climate change. Trends Plant Sci 2010; 15(12): 664-74.
[http://dx.doi.org/10.1016/j.tplants.2010.08.002] [PMID: 20846898]

[4]   Sassi-Audi S, Audi S, Abdelly C. Inoculation with the native *Rhizobium gallicum* 8a3 improves osmotic stress tolerance in common bean drought-sensitive cultivar. Soil Plant Sci 2012; 62: 179-87.

[5]   Fernandez-Laqueno F, Dendooven L, Munive A, Corlay-Chee L, Serrano-Covarrubias LM, Espinosa-Victor D. Micro-morphology of common bean (*Phaseolus vulgaris* L.) nodules undergoing senescence. Acta Physiol Plant 2008; 30: 545-52.
[http://dx.doi.org/10.1007/s11738-008-0153-7]

[6]   Vorster BJ, Schluter U, du Plessis M, *et al.* The cysteine protease cysteine-proteases inhibitor system explored in soybean nodule development. Agron Dis 2013; 3: 550-70.
[http://dx.doi.org/10.3390/agronomy3030550]

[7]   van Wyk SG, Du Plessis M, Cullis CA, Kunert KJ, Vorster BJ. Cysteine protease and cystatin expression and activity during soybean nodule development and senescence. BMC Plant Biol 2014; 14: 294-9.
[http://dx.doi.org/10.1186/s12870-014-0294-3] [PMID: 25404209]

[8]   Oehrle NW, Sarma AD, Waters JK, Emerich DW, Emerich DW. Proteomic analysis of soybean nodule cytosol. Phytochemistry 2008; 69(13): 2426-38.
[http://dx.doi.org/10.1016/j.phytochem.2008.07.004] [PMID: 18757068]

[9]   Crews TE, Peoples MB. Legumes *versus* fertilizers sources of nitrogen: ecological trade-offs and human needs. J Agric Eco Environ 2004; 102: 279-97.
[http://dx.doi.org/10.1016/j.agee.2003.09.018]

[10]   McDonald AJ, Powlson DS, Poulton PR, Jenkinson DS. Unused fertilizer nitrogen in arable soils—its contribution to nitrate leaching. J Sci Food Agric 2006; 46: 407-19.
[http://dx.doi.org/10.1002/jsfa.2740460404]

[11]   Li D, Yee JA, McGuire MH, Murphy PA, Yan L. Soybean isoflavones reduce experimental metastasis in mice. J Nutr 1999; 129(5): 1075-8.
[http://dx.doi.org/10.1093/jn/129.5.1075] [PMID: 10222402]

[12]   Messina MJ. Legumes and soybean: An overview of their nutritional profiles and health effects. J Clin Nutr 1999; 10: 439-50.
[http://dx.doi.org/10.1093/ajcn/70.3.439s]

[13]   Mateos-Aparicio I, Redondo Cuenca A, Villanueva-Suárez MJ, Zapata-Revilla MA. Soybean, a promising health source. Nutr Hosp 2008; 23(4): 305-12.
[PMID: 18604315]

[14]   Dakora FD, Keya SO. Contribution of legume nitrogen fixation to sustainable agriculture in Sub-Saharan Africa. Soil Biol Biochem 1997; 29: 809-17.
[http://dx.doi.org/10.1016/S0038-0717(96)00225-8]

[15]   Peoples MB, Bowman AM, Gault RR, *et al.* Factors regulating the contributions of fixed nitrogen by pasture and crop legumes to different farming systems in eastern Australia. Plant Soil 2001; 228: 29-41.
[http://dx.doi.org/10.1023/A:1004799703040]

[16]   Poeples MB, Herridge DF, Ladha JK. Biological nitrogen fixation: an efficient source of nitrogen for sustainable agricultural production. Plant Soil 1995; 174: 3-28.
[http://dx.doi.org/10.1007/BF00032239]

[17]   Ashraf M, Foolad MR. Roles of glycine betaine and proline in improving plant abiotic stress resistance. Environ Exp Bot 2007; 59: 206-16.
[http://dx.doi.org/10.1016/j.envexpbot.2005.12.006]

[18]   Park C, Park JM. Endoplasmic reticulum plays a critical role in integrating signals generated by both biotic and abiotic stress in plants. Front Plant Science 2019.
[http://dx.doi.org/10.3389/fpls.2019.00399]

[19]   Afrin T, Diwan D, Sahawneh K, Pajerowska-Mukhtar K. Multilevel regulation of endoplasmic reticulum stress responses in plants: where old roads and new paths meet. J Exp Bot 2020; 71(5): 1659-67.
[http://dx.doi.org/10.1093/jxb/erz487] [PMID: 31679034]

[20]   Valente MAS, Faria JAQA, Soares-Ramos JR, *et al.* The ER luminal binding protein (BiP) mediates an increase in drought tolerance in soybean and delays drought-induced leaf senescence in soybean and tobacco. J Exp Bot 2009; 60(2): 533-46.
[http://dx.doi.org/10.1093/jxb/ern296] [PMID: 19052255]

[21]   Alam I, Sharmin SA, Kim K, *et al.* Comparative proteomic approach to identify proteins involved in flooding combined with salinity stress in soybean. An Int J Plant-Soil Rel 2011; 346: 45-62.
[http://dx.doi.org/10.1007/s11104-011-0792-0]

[22]   Arai Y, Hayashi M, Nishimura M. Proteomic analysis of highly purified peroxisomes from etiolated soybean cotyledons. Plant Cell Physiol 2008; 49(4): 526-39.
[http://dx.doi.org/10.1093/pcp/pcn027] [PMID: 18281324]

[23]   Hoa TP, Nomura M, Kajiwara H, Day DA, Tajima S. Proteomic analysis on symbiotic differentiation of mitochondria in soybean nodules. Plant Cell Physiol 2004; 45(3): 300-8.
[http://dx.doi.org/10.1093/pcp/pch035] [PMID: 15047878]

[24]   Panter S, Thomson R, de Bruxelles G, Laver D, Trevaskis B, Udvardi M. Identification with proteomics of novel proteins associated with the peribacteroid membrane of soybean root nodules. Mol Plant Microbe Interact 2000; 13(3): 325-33.
[http://dx.doi.org/10.1094/MPMI.2000.13.3.325] [PMID: 10707358]

[25]   Xu C, Sullivan JH, Garrett WM, Caperna TJ, Natarajan S. Impact of solar ultraviolet-B on the proteome in soybean lines differing in flavonoid contents. Phytochemistry 2008; 69(1): 38-48.
[http://dx.doi.org/10.1016/j.phytochem.2007.06.010] [PMID: 17645898]

[26]   Zhen Y, Qi JL, Wang SS, *et al.* Comparative proteome analysis of differentially expressed proteins induced by Al toxicity in soybean. Physiol Plant 2007; 131(4): 542-54.
[http://dx.doi.org/10.1111/j.1399-3054.2007.00979.x] [PMID: 18251846]

[27]   Aghaei K, Ehsanpour AA, Shah AH, Komatsu S. Proteome analysis of soybean hypocotyl and root under salt stress. Amino Acids 2009; 36(1): 91-8.
[http://dx.doi.org/10.1007/s00726-008-0036-7] [PMID: 18264660]

[28]   Shi F, Yamamoto R, Shimamura S, *et al.* Cytosolic ascorbate peroxidase 2 (cAPX 2) is involved in the soybean response to flooding. Phytochemistry 2008; 69(6): 1295-303.
[http://dx.doi.org/10.1016/j.phytochem.2008.01.007] [PMID: 18308350]

[29]   Sobkowiak R, Deckert J. Proteins induced by cadmium in soybean cells. J Plant Physiol 2006; 163(11): 1203-6.
[http://dx.doi.org/10.1016/j.jplph.2005.08.017] [PMID: 17032622]

[30]   Mohammadi PP, Moieni A, Hiraga S, Komatsu S. Organ-specific proteomic analysis of drought-stressed soybean seedlings. J Proteomics 2012; 75(6): 1906-23.
[http://dx.doi.org/10.1016/j.jprot.2011.12.041] [PMID: 22245419]

[31]   Van de Velde W, Guerra JC, De Keyser A, *et al.* Aging in legume symbiosis. A molecular view on nodule senescence in *Medicago truncatula.* Plant Physiol 2006; 141(2): 711-20.
[http://dx.doi.org/10.1104/pp.106.078691] [PMID: 16648219]

[32]   Smith PK, Krohn RI, Hermanson GT, *et al.* Measurement of protein using bicinchoninic acid. Anal Biochem 1985; 150(1): 76-85.
[http://dx.doi.org/10.1016/0003-2697(85)90442-7] [PMID: 3843705]

[33]   Verma S, Dixit R, Pandey KC. Cysteine Proteases: Models of activation and future prospects as pharmacological targets. Front Pharmacol 2016; 7: 107.
[http://dx.doi.org/10.3389/fphar.2016.00107] [PMID: 27199750]

[34]   Beers EP, Woffenden BJ, Zhao C. Plant proteolytic enzymes: possible roles during programmed cell death. Plant Mol Biol 2000; 44(3): 399-415.
[http://dx.doi.org/10.1023/A:1026556928624] [PMID: 11199397]

[35]   Alesandrini F, Mathis R, Van de Sype G, Hêrouart D, Puppo A. Possible roles for a cysteine protease and hydrogen peroxide in soybean nodule development and senescence. New Phytol 2003; 158: 131-8.
[http://dx.doi.org/10.1046/j.1469-8137.2003.00720.x]

[36]   Solomon M. belenghi B, Delledonne M, Menachem E, Levine A. The involvement of cysteine proteases and protease inhibitor genes in the regulation of programmed cell death in plants. J Plant Cell 1999; 11: 43-443.

[37]   Azarkan M, El Moussaoui A, van Wuytswinkel D, Dehon G, Looze Y. Fractionation and purification of the enzymes stored in the latex of *Carica papaya*. J Chromatogr B Analyt Technol Biomed Life Sci 2003; 790(1-2): 229-38.
[http://dx.doi.org/10.1016/S1570-0232(03)00084-9] [PMID: 12767335]

[38]   Martínez M, Cambra I, González-Melendi P, Santamaría ME, Díaz I. C1A cysteine-proteases and their inhibitors in plants. Physiol Plant 2012; 145(1): 85-94.
[http://dx.doi.org/10.1111/j.1399-3054.2012.01569.x] [PMID: 22221156]

[39]   Wiederanders B. Structure-function relationships in class CA1 cysteine peptidase propeptides. Acta Biochim Pol 2003; 50(3): 691-713.
[http://dx.doi.org/10.18388/abp.2003_3661] [PMID: 14515150]

[40]   Cilliers M, van Wyk SG, van Heerden PDR, Kunert KJ, Vorster BJ. Identification and changes of the drought-induced cysteine protease transcriptome in soybean (*Glycine max*) root nodules. Environ Exp Bot 2018; 148: 59-69.
[http://dx.doi.org/10.1016/j.envexpbot.2017.12.005]

[41]   Hara-Nishimura I, Hatsugai N, Nakaune S, Kuroyanagi M, Nishimura M. Vacuolar processing enzyme: an executor of plant cell death. Curr Opin Plant Biol 2005; 8(4): 404-8.
[http://dx.doi.org/10.1016/j.pbi.2005.05.016] [PMID: 15939660]

[42]   Li Y, Zhou L, Li Y, *et al.* A nodule-specific plant cysteine proteinase, AsNODF32, is involved in nodule senescence and nitrogen fixation activity of the green manure legume *Astragalus sinicus*. New Phytol 2008; 180(1): 185-92.
[http://dx.doi.org/10.1111/j.1469-8137.2008.02562.x] [PMID: 18643938]

[43]   Drake R, John I, Farrell A, Cooper W, Schuch W, Grierson D. Isolation and analysis of cDNAs encoding tomato cysteine proteases expressed during leaf senescence. Plant Mol Biol 1996; 30(4): 755-67.
[http://dx.doi.org/10.1007/BF00019009] [PMID: 8624407]

[44]   Vorster BJ, Cullis CA, Kunert KJ. Plant vacuolar processing enzymes. Front Plant Sci 2019; 10: 479.
[http://dx.doi.org/10.3389/fpls.2019.00479] [PMID: 31031794]

[45]   Mangena P. Phytocystatins and their potential application in the development of drought tolerance plants in soybeans (*Glycine max* L.). Protein Pept Lett 2020; 27(2): 135-44.
[http://dx.doi.org/10.2174/0929866526666191014125453] [PMID: 31612812]

[46]   Oh CJ, Lee H, Kim HB, An CS. Isolation and characterization of a root nodule-specific cysteine proteinase cDNA from soybean. J Plant Biol 2004; 47.
[http://dx.doi.org/10.1007/BF03030511]

[47]   Kunert KJ, Vorster BJ, Fenta BA, Kibido T, Dionisio G, Foyer H. Drought stress responses in soybean roots and nodules. Front Plant Sci 2016; (7): 1015.
[http://dx.doi.org/10.3389/fpls.2016.01015]

[48]   Otegui MS, Noh YS, Martínez DE, *et al.* Senescence-associated vacuoles with intense proteolytic activity develop in leaves of *Arabidopsis* and soybean. Plant J 2005; 41(6): 831-44.
[http://dx.doi.org/10.1111/j.1365-313X.2005.02346.x] [PMID: 15743448]

[49]   Kardailsky IV, Brewin NJ. Expression of cysteine protease genes in pea nodule development and senescence. Amer Phytopathol Soc 1996; 9: 689-95.
[http://dx.doi.org/10.1094/MPMI-9-0689]

[50]   Lidgett AJ, Moran M, Wong KA, Furze J, Rhodes MJ, Hamill JD. Isolation and expression pattern of a cDNA encoding a cathepsin B-like protease from Nicotiana rustica. Plant Mol Biol 1995; 29(2): 379-84.
[http://dx.doi.org/10.1007/BF00043660] [PMID: 7579187]

[51]   Benchabane M, Schlüter U, Vorster J, Goulet MC, Michaud D. Plant cystatins. Biochimie 2010; 92(11): 1657-66.
[http://dx.doi.org/10.1016/j.biochi.2010.06.006] [PMID: 20558232]

[52]   Turk B, Turk D, Salvesen GS. Regulating cysteine protease activity: essential role of protease inhibitors as guardians and regulators. Curr Pharm Des 2002; 8(18): 1623-37.
[http://dx.doi.org/10.2174/1381612023394124] [PMID: 12132995]

[53]   Chu MH, Liu KL, Wu HY, Yeh KW, Cheng YS. Crystal structure of tarocystatin-papain complex: implications for the inhibition property of group-2 phytocystatins. Planta 2011; 234(2): 243-54.
[http://dx.doi.org/10.1007/s00425-011-1398-8] [PMID: 21416241]

[54]   Nagata K, Kudo N, Abe K, Arai S, Tanokura M. Three-dimensional solution structure of oryzacystatin-I, a cysteine proteinase inhibitor of the rice, *Oryza sativa* L. japonica. Biochemistry 2000; 39(48): 14753-60.
[http://dx.doi.org/10.1021/bi0006971] [PMID: 11101290]

[55]   Anastasi A, Brown MA, Kembhavi AA, *et al.* Cystatin, a protein inhibitor of cysteine proteinases. Improved purification from egg white, characterization, and detection in chicken serum. Biochem J 1983; 211(1): 129-38.
[http://dx.doi.org/10.1042/bj2110129] [PMID: 6409085]

[56]   Turk V, Bode W. The cystatins: protein inhibitors of cysteine proteinases. FEBS Lett 1991; 285(2): 213-9.
[http://dx.doi.org/10.1016/0014-5793(91)80804-C] [PMID: 1855589]

[57]   van Wyk SG, Kunert KJ, Cullis CA, *et al.* Review: The future of cystatin engineering. Plant Sci 2016; 246: 119-27.
[http://dx.doi.org/10.1016/j.plantsci.2016.02.016] [PMID: 26993242]

[58]   Botella MA, Xu Y, Prabha TN, *et al.* Differential expression of soybean cysteine proteinase inhibitor genes during development and in response to wounding and methyl jasmonate. Plant Physiol 1996; 112(3): 1201-10.
[http://dx.doi.org/10.1104/pp.112.3.1201] [PMID: 8938418]

[59]   Quain MD, Makgopa ME, Cooper JW, Kunert KJ, Foyer CH. Ectopic phytocystatin expression increases nodule numbers and influences the responses of soybean (*Glycine max*) to nitrogen deficiency. Phytochemistry 2015; 112: 179-87.
[http://dx.doi.org/10.1016/j.phytochem.2014.12.027] [PMID: 25659749]

[60]   Prins A, van Heerden PDR, Olmos E, Kunert KJ, Foyer CH. Cysteine proteinases regulate chloroplast protein content and composition in tobacco leaves: a model for dynamic interactions with ribulose-1,--bisphosphate carboxylase/oxygenase (Rubisco) vesicular bodies. J Exp Bot 2008; 59(7): 1935-50.
[http://dx.doi.org/10.1093/jxb/ern086] [PMID: 18503045]

[61]   Zhang X, Liu S, Takano T. Two cysteine proteinase inhibitors from Arabidopsis thaliana, AtCYSa and AtCYSb, increasing the salt, drought, oxidation and cold tolerance. Plant Mol Biol 2008; 68(1-2): 131-43.
[http://dx.doi.org/10.1007/s11103-008-9357-x] [PMID: 18523728]

[62]   Yuan S, Li R, Wang L, *et al.* Search for nodulation and nodule development-related cystatin genes in the genome of soybean (*Glycine max*). Front Plant Sci 2016.

CHAPTER 7

Cowpea Production, Uses and Breeding

Phetole Mangena[*], **Erlafrida Ramokgopa** and **Lifted Olusola**

Department of Biodiversity, School of Molecular and Life Sciences, Faculty of Science and Agriculture, University of Limpopo, Private Bag X1106, Sovenga 0727, South Africa

Abstract: Cowpea, *Vigna unguiculata* L. is a very important grain legume crop that is grown in the tropic and sub-tropical regions. It provides a strong support to the livelihood of the rural poor people and small scale farmers through contributions to their nutritional security, income generation and the improvement of soil fertility. However, its production yield can be adversely affected by abiotic and biotic constraints. The stress affecting cowpea creates the need to develop and implement breeding strategies that can alleviate the devastations caused by biotic and abiotic constraints. Breeders employ pedigree, backcross, marker-assisted breeding, genome editing (CRISPR-Cas9) and other modern biotechnological techniques for genetic manipulation of cowpeas, including legumes such as soybean, chickpea and common bean. These useful strategies have brought about major opportunities for breeders to develop cowpea cultivars with improved tolerance to a wide range of growth and yield inhibiting stress factors.

Keywords: Breeding, Cowpea, Improvement, Productivity, Stress constraints.

INTRODUCTION

Vigna unguiculata L., commonly known as the "cowpea", is a member of the legumes, belonging to the Fabaceae family. This group of dicotyledonous plants comprises of trees, shrubs, and herbs with both erect and climbing growth tendency. Although, the cowpea only has considerable amounts of methionine and cysteine, they are a valuable source of essential amino acids, mineral iron, copper, and zinc. They supply proteins, carbohydrates, essential elements, and vitamins used as foods for humans and feeds for animals [1].

[*] **Corresponding author Phetole Mangena**: Department of Biodiversity, School of Molecular and Life Sciences, Faculty of Science and Agriculture, University of Limpopo, Private Bag X1106, Sovenga 0727, South Africa; Tel: +2715-268-4715; E-mails: mangena.phetole@gmail.com & phetole.mangena@ul.ac.za

Due to the significant amount of dietary proteins and the fact that it serves as a supplement for amino acid in poor diets, it is consumed worldwide, making it one of the most commonly consumed legumes in some regions with a high commercial value and a worldwide yield average of approximately over 700 kg/ha, almost similar to common bean [2].

Grain legumes, like cowpea, are cultivated mostly in regions that have little or no agricultural or industrial value, and they rely primarily on rainwater rather than irrigation. Fortunately, this crop exhibits a significant level of abiotic stress tolerance (heat and drought), in contrast to the most damaging biotic stress factors. The successful plant growth and development in cowpea usually occur under normal environmental conditions wherein the mineral elements, soil water potential, temperature and other plant requirements are a little less within their optimum range [1]. However, plants often encounter unusual and extreme conditions in their habitats. Grain legumes have proved generally sensitive to environmental extremes, whereby their exposure to prolonged feverish temperatures and limited soil moisture severely cause major losses in yields, particularly in the tropics and sub-Saharan African (SSA) region.

In the SSA, including other tropical regions like Central America and India, these conditions are exacerbated by frequent droughts and salinity stress that are often caused by the effects of climate change [3, 4]. The extremes in environmental parameters create stressful conditions for all plants, which may have severe impacts on their physiological processes, growth, overall development, and survival. Thus, the gap between production levels and the demand for improved grain legume cultivars that withstand harmful growth conditions continues to widen. This also remains the case for cowpea due to the limited genetic diversity available for breeding programs, as well as the increasingly unpredictable global climates. However, cowpea already exhibits a considerable level of abiotic stress tolerance, especially for water deficit, but its exposure to these severe kinds of stress accelerates the spread and attack of cowpea by insect pests as well as fungal, bacterial and virus diseases.

To alleviate the devastations caused by cowpea production constraints, breeding systems involving technologies such as marker-assisted selection, genome editing and genetic transformation *via* biolistics/ electroporation/ bio-transformation (using *Agrobacterium* spp.) are being investigated. The integration of both novel molecular and conventional breeding techniques is routinely tested for the development of improved cowpea lines with high yield potential and resistance. Therefore, this chapter will provide a succinct review on cowpea production, uses and breeding strategies of the reported genotypes, and further discusses novel ways in which these crops have been genetically enhanced for resistance against biotic and abiotic stress factors. This chapter will also examine some of the biotic stresses that cowpea plants commonly encounter in their environment. The basic concepts of biological stress, acclimation and adaptation will be briefly analysed in conjunction with the response of cowpea to insect attacks, diseases (fungal, bacterial and/or viral), as well as other environmental stress factors.

IMPORTANCE OF COWPEA

Cowpea is a grain legume that is grown globally in dry areas of the tropics and the subtropics. It plays a very important role in human nutrition and food security as well as for generating income for farmers and food vendors [3]. The grains or dry seeds are the most important part of the plant used for human consumption [4, 5]. Cowpea has protein-rich grains which also contain carbohydrates, folic acid and some minerals and vitamins. In the Eastern and Southern Africa, the young leaves of cowpea are consumed as spinach, while the green immature and green mature seed pods are also prepared as a vegetable stew (Fig. **1**), especially in Senegal, South Africa and other African countries [3]. The evaluation of 1541 germplasm lines revealed that cowpea grains on average contain 24% protein, 53.2 mg/kg iron, 38.1 mg/kg zinc, 826 mg/kg calcium, 1915 mg/kg magnesium, 14,890 mg/kg potassium and 5055 mg/kg phosphorus [3, 5].

The nutritional profile of cowpea grain is comparable to that of other pulses with a relatively low-fat content and a total protein as well as carbohydrate content that is two to four-folds higher than cereal and tuber crops (Fig. **1**) [6]. Furthermore, the biomass from cowpea plants provides important nutritious fodder for ruminants, and this has been reported by Boukar *et al.* [3] as the most common practice in Sahel regions, as well as in other parts of western and central Africa. Similar to other leguminous grain/pulse crops, cowpea has the ability to fix atmospheric nitrogen, which further contributes to soil fertility. Cowpea fixes between 70 and 350 kg of nitrogen per hectare and contributes more than 80 kg of nitrogen per hectare in the soil [7].

However, the nitrogen fixation capabilities of *Vigna* spp., including the *Phaseoleae* spp., usually vary according to the genotype, and it is generally less than that of other prominent and agronomically important legumes such as soybean (*Glycine max* L.). Soybean has a larger frequency of root nodulation, accompanied by a higher nitrogenase activity. The high rates of nitrogen fixation still take place in both inoculated and uninoculated varieties. However, the lesser N fixation capacity observed in other legume crops such as cowpea and common bean may be due to a wide range of native and ineffective *Rhizobium* spp (Fig. **1**) that heavily colonise the roots while some microbial cells take out too little or no nitrogenase activity [8]. Soybean fixes more than 160 kg of atmospheric N per ha into the soil [9] compared to 116, 87 and 74 kg N ha^{-1} for Ife BPC, Ife Brown and AFB 1757 obtained in cowpea treated with different strains of *Bradyrhizobium* [10]. Additionally, Karoney *et al.* [11] postulated that root infection by nitrogen-fixing *Rhizobium* also triggers enzyme-mediated induced resistance reactions, which may lead to the production of defensive compounds that suppress aboveground colonisation by foliar pests. But, the results indicated that *Rhizobium* promotes host plant suitability for *Colletotrichum lindemuthianum*, and possibly enhances host plant tolerance to the pathogen instead of conferring complete resistance.

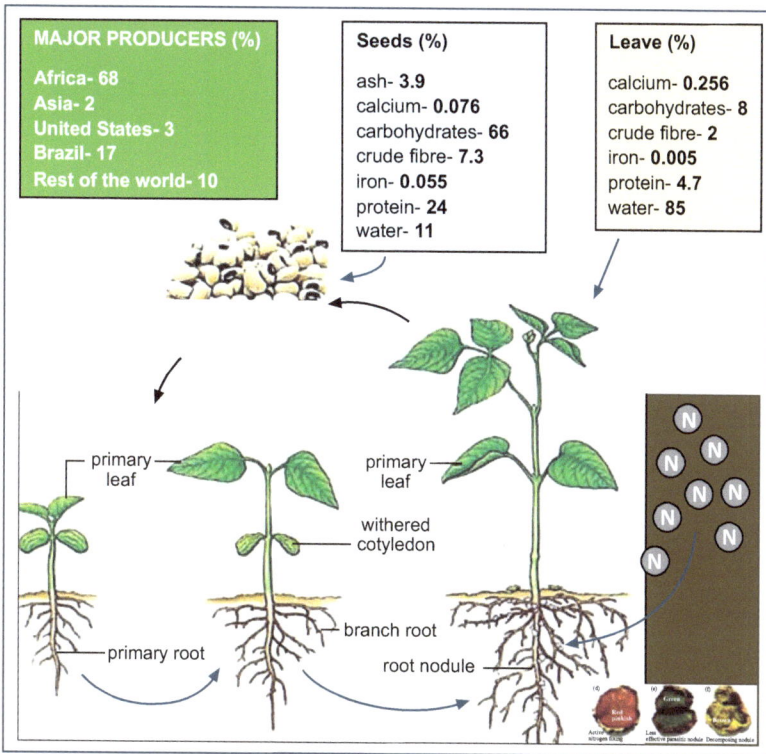

Fig. (1). Cowpea growth stages from seedling to adult plant, overall world production stats (2017-2018), nodulation and chemical compositions in seeds and leaves harvested for consumption [1, 12].

DISTRIBUTION AND PRODUCTION

Although accurate statistics are generally unavailable, obtainable records on cowpea production are estimated at about 6.4 million metric tons of grains annually. This figure is produced from a total cultivated area of 14.5 million hectares worldwide [12]. Approximately 83% of the global cowpea production is cultivated in Africa, with over 80% of this production taking place in West Africa [13]. It was estimated that in Africa, Nigeria contributes up to 55% of cowpea production and 45% worldwide, which makes it the world's largest producer and consumer of cowpea, followed by the Republic of Niger with 15%, 12% in Brazil and 5% in Burkina Faso [12]. Table **1** below shows the top 12 producers of cowpea in West and Central Africa between 1999 and 2018. In the last three decades, the global production of cowpea has grown at an average of 5% annual growth rate, with 3.5 increase in the production area and about 1.5% improvement in yield.

Table 1. Brief estimates of cowpea harvested area, average yield and the total (%) production in West and Central African countries.

	Harvested Area (m.ha)	Average Yield (kg.ha⁻¹)	Total Production (%)
Nigeria	5.8	530	47.2
Niger	4.1	325	20.0
Mali	2.22	244	7.9
Burkina Faso	2.01	218	15.6
Togo	1.35	284	3.8
Benin	1.0	521	6.4
Senegal	9.5	341	3.2
Ghana	8.5	663	5.7
Mauritania	5.2	331	1.7
Cote d' Ivoire	4.0	500	2.0
Chad	4.4	489	2.1
Cameroon	3.8	827	3.1

Sources: FAO [12], Langyintuo *et al.* [14], Carsky [15], Sanou *et al.* [16], Saka *et al.* [17].
m.ha- million hectares, **kg.ha⁻¹**- kilogram per hectare, **%**- percent.

Area expansion of production accounted for 70% of the total growth during this period [18, 19], and the global share of cowpea based on the total area under pulses grew from less than 10 to 20% between 1990 and 2007. Currently, cowpea occupies over 85% of the area under pulses cultivation and 10% of the total cultivated land in West Africa [19]. The global cowpea supply is predicted to reach 9.8 million metric tons in 2030 if the trends in cowpea area expansion and yield keep increasing with time. Therefore, there is a need for increased investment in research, particularly in order to increase the growth and yield of cowpea to meet the predicted demands, as well as preventing any possible deficits [20]. Furthermore, cowpea production and consumption do not occur simultaneously; farmers, traders and consumers also need efficient long-term storage systems to ensure continued grain availability for consumers, as indicated by Langyintuo *et al*. [14]. Cowpea has the greatest potential to serve as one of the world's richest sources of vitamins, mineral nutrients, energy, and relatively low-cost proteins for under-developed countries.

STRESS RESISTANCE

The various biotic and abiotic stress effects limit the production of cowpea. Major culprits are the biotic stresses, which include insect pests, diseases, parasitic weeds and nematode, demonstrated in Table **2** and Fig. (**2**). However, cowpea exhibits a considerable level of resistance to abiotic stress. According to Boukar *et al*. [21], abiotic stress constraints that also limit the growth and production of cowpea may include prolonged and severe drought, heat stress, chilling stress and low soil fertility. These environmental constraints, together with diseases/pests, are responsible for reducing crop yield and grain quality to very minimal figures or even cause complete losses (90–100%), depending on factors such as resource availability to farmers, season, geographical location and intensity of stress [22]. For example, the average yield losses could range between 50 to 100% on pesticide-untreated legume grown fields.

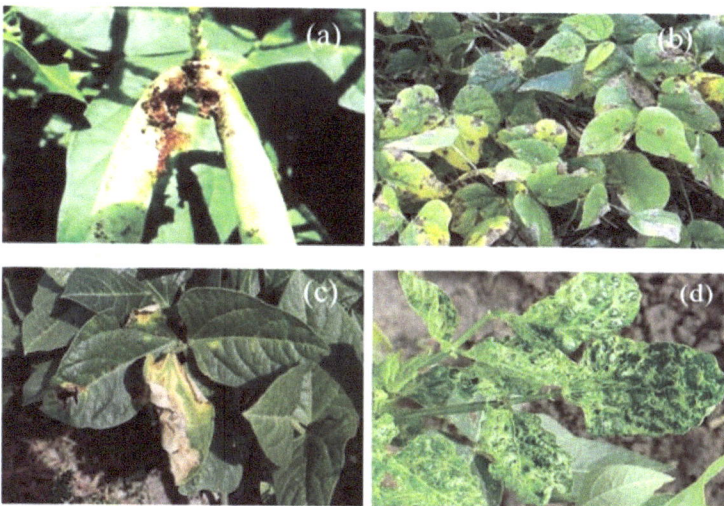

Fig. (2). Insect-cowpea pod borer (**a**), Cowpea fungal disease (**b**), Cowpea bacterial wilt (**c**), and Cowpea severe mosaic virus (**d**) [27, 28, 33].

Since cowpea exhibits the inherent capacity to tolerate abiotic stress, this resistance still requires that the plants maintain thermodynamic equilibrium with the stress [1], ensuring that the outside environmental conditions do not severely interfere with internal plant growth and developmental processes. Drought tolerance, for example, requires that plants survive water deficit stress without sustaining any major protoplast injuries. Similarly, for biotic stress occurring under the same natural environment, plants that are exposed to a variety of potentially pathogenic microorganisms should be capable of presenting a non-host resistance. This type of resistance has been described by Fink *et al.* [27] to involve complex genetics with a considerable number of genes that may be able to cause diseases in one or few plant species but usually fail to be pathogenic in most others.

Table 2. Summary of some of the principal biotic stress factors found in cowpea plants and their descriptions, as well as their probable effects.

Description and Type of Insect Pests	Sources
There are three major groups of cowpea pests attacks based on the time of infestation relative to crop phenology (from seedling to harvest): - Leaf hoppers, cowpea aphids and foliage beetles are pests which are commonly found throughout the vegetative growth of the crop; they are vectors to viruses and they also feed on leaves, - Flower bud thrips, lepidopterous larvae and beetles are pests which infest crop flowers based on the appearance of flowers and cause flower abortion and destroy buds, - Legume pod borer, a complex of pod sucking buds, and the storage weevil are pests which are prevalent throughout the reproductive period and storage.	Boukar *et al.* [3], Boukar and Fatokun [22], Jackai and Daoust [23], Singh *et al.* [24], Singh and Van Emden [25], and Singh and Allen [26].
Description and Type of Diseases	
Cowpea can be affected by numerous bacterial, fungal and viral diseases, which results in intensive yield reductions. The severity of the disease is influenced by the region even though some diseases are found worldwide and cause consistent damage to the crop. ***Fungal Diseases*** - Pre-emergence and post-emergence damping off caused by *Pythium altimum*, - *Fusarium* wilt caused by *Fusarium oxysporium*, - *Macrophomina* blight caused by *Macrophomina phaseolina*, - Web blight and root rot caused by *Phytophthora vignae*, - Scab caused by *Sphacelona sp.*, - Leaf spot caused by *Pseudocercospora crueta* and *Cercospora apii*. ***Bacterial Diseases*** - Bacterial blight caused by *Xanthomonas axonopodis* pv. *Phaseoli*. Bacterial blight causing stem canker under serious infection, which could eventually kill the affected plant, - Bacterial pustule caused by *Xanthomonas sp.* occurs sporadically and is less damaging to cowpea than bacterial blight. ***Viral Diseases*** - A range of severe mosaics caused by Cowpea Mosaic Virus, - An inconspicuous green mottle to severe mosaic, leaf distortion, blistering and death of the plant caused by Cowpea (yellow) Mosaic Virus, - Various mosaics and mottling caused by Cowpea Aphid-borne Mosaic Virus, - Stunted yellow plants, distorted leaves and blistering caused by Cowpea Golden Mosaic	

Biotic Stress

Insect pests are major biotic constraints in cowpea production, causing total yield failure in cases of severe plant attacks. Often several pests (Table 2 and Fig. 2) may attack a plant at a time and cause yield losses at every stage of the cowpea's life cycle [3, 28]. With pests and plant diseases being a continual problem for agriculture, due to the genetic adaptability of the pathogens, this problem requires a sustainable strategy to develop disease-resistant varieties. For example, *Gloeosporium* spp. (*G. lindemuthianum*) has been reported to affect yield and seed quality of cowpea. The fungus infects all growth and reproductive stages of the plant, such as seedling development, vegetative growth, flowering and podding, particularly the leaves, pods and seeds. These fungal pathogens mechanically penetrate the cuticle and epidermal cells, internally enlarging their hyphae and colonising tissues along the plant's cell walls and protoplasts [29]. Viral and bacterial infections, on the other hand, target leaf and vascular tissues to spread the infection in order to cause mottling, necrotic/chlorotic spots, stunting and curling of leaves and stems [1, 22].

Nsa and Kareem [30] reported viral symptoms that included apical necrosis, defoliation, mottling and leaf reduction in three cowpea cultivars (commercial white cultivar and two IITA lines: IT8ID-985 and TVu76) using mixed viruses infections. An emerging seedborne bacterial wilt and tan spot caused by *Curtobacterium flaccumfaciens* were also reported by Osdaghi *et al.* [31]. This bacterium was reported in Osdaghi *et al.*'s study to cause various seed decoloration (yellow, orange, pink or purple) in white seeded cultivars, as well as interveinal chlorosis, necrosis and systemic wilt on new and older leaflets in cowpea, common bean, mung bean and soybean. Although plants produce phytoalexins to help ward-off insects and diseases, the sector still utilises cultural and chemical control to avert large-scale yield losses for growers. Several strategies are currently being used to eliminate pests shown in Fig. (2) and Table 1, and to produce pathogen-free and resistant plants.

As the extent of insect pests' infestation and disease continues to expand, some researchers, including Togola *et al.* [32], believe that the application of synthetic insecticides remains the most commonly used strategy for combating pests attacks in cowpea. This approach is still very popular despite its negative impact on the environment, human and animal health. A report by Mohammed [29] suggested a few traditional methods of eradicating and reducing the spread of pathogens merely through good management practices of seed-borne diseases. The review recommended complete removal of infested debris immediately after harvest, soil solarization using transparent plastic for sheeting, done at least a month before sowing, and/or seed pretreatment with bio-stimulants, such as *Trichoderma* spp.

liquid inoculum to enhance the crop's growth performance. Other efficient and highly effective methods include the use of resistant cultivars, especially for plant breeding purposes, and integrated management methods that involve the use of botanical as well as biopesticides along with fungicides for pests control.

Another improved integrated management strategy for the cultivation of cowpea was reported by Chikutuma *et al.* [34] with the aim of promoting conservation agriculture. This study reported reduced herbicide effects on the environment when using atrazine, especially by mixing with other herbicides such as glyphosate.

Abiotic Stress

The traditional agricultural practices such as reducing soil disturbances (tilling), maintaining soil cover through residue retention, use of green manure and crop rotation systems have always been used to achieve high cowpea yields and good grain quality. However, abiotic stresses that include drought, heat, chilling, low soil fertility, *etc.* also influence cowpea growth and productivity. These factors are often not efficiently combatted, especially on bigger agricultural fields compared to biotic stress that can be eradicated through biopesticides and chemical control. Although, cowpea is known to be a drought-tolerant crop, its yield can be reduced significantly when plants are exposed to mid-season or terminal drought during seedling development and reproductive stages. Serious damage may be caused by heat during the off-season cropping and high night temperatures, which may cause flower abortion, thus preventing pod formation and leading to consequent reduction of grain yield. According to Boukar *et al.* [3] and Namugwanya *et al.* [18], soils that lack nutrients, for example phosphorus deficiency, can negatively affect nitrogen fixation in cowpea root nodules and afterward lower the productivity potential of the crop. Inherent resistance to abiotic stress provides a genetic basis and resources to advance breeding programs aimed at producing newly improved cowpea cultivars. This section onward briefly highlights the effects of some of the major abiotic constraints and then discusses various strategies in which these factors, together with disease-causing pathogens and pests, could be combatted.

Effects of Drought Stress on Cowpea

Frequent drought causes a considerable amount of damage in many leguminous crops, including cowpea (Fig. **3**) [33]. Growers in tropical and subtropical regions such as Africa Sahelian zones and the northeast region of Brazil cultivate cowpea only during the rainy seasons. This indicates how drought represents the most important threat to growth, yield and biomass production, particularly for many small farmers in developing countries [35]. Consequently, the cultivation of legume crops under drought-prone areas would require various factors and mechanisms that could work independently or jointly to enable plants to tolerate drought stress [36]. Drought tolerance can be traditionally defined as the ability of plants to live, grow and yield satisfactorily with limited soil water supply or under periodic water deficiencies [37]. Cowpea plants can avoid lethal water potential and other water deficits by preventing excessive water loss through the stomata. Many plants, including C_3 and CAM, use physiological traits like stomatal control to regulate cell water potential [1]. Bastos *et al.* [38] and Nascimento *et al.* [39] reported that cowpea plants can produce more than 1000 kg grain ha^{-1} but the drought stress can reduce such potential to approximately 360 kg ha^{-1} before flowering. Reduced precipitation and long drought periods are still being predicted, especially as a result of climate change [40].

Effects of Heat Stress on Cowpea

Heat damages all plant processes irreversibly if plants are exposed to high temperatures for longer durations. High temperatures above 35°C cause flower abortion by affecting pollen and stigma development. These effects eventually cause cowpea yield losses of sensitive varieties by preventing seed formation and pod set [41]. Furthermore, Warrag and Hall [42] reported that high night temperatures can be much more detrimental to grain yield than high temperatures during the day. This study further showed that high night temperatures cause androecium sterility and reduce grain yield by increasing floral abscission and decreasing the number of pods harvested. Although plants exhibit a wide range of sensitivities to extreme temperatures, some are killed or injured by moderate heat and chilling temperature while those that acclimatised could survive certain temperature conditions (Fig. **3**) [1, 41]. However, with the constant global warming and the uncertainty of the effects of climate change, legumes and other crops will continue experiencing increased temperatures that can adversely affect their growth and yield.

Effects of Low Soil Fertility on Cowpea

Low soil fertility as a result of low organic matter and low phosphorus in the Savannah is a major production constraint for cowpea (Fig. **3**) [21]. Phosphorus (P) has been reported by Karikari *et al.* [43] as a major limiting nutrient that negatively affects growth, nodulation and yield in cowpea. Phosphorus is critically involved in plant processes such as energy metabolism, nitrogen fixation, nucleic acids synthesis, photosynthesis, respiration, synthesis of cell membranes and enzyme regulations. In addition to P, cowpea production in such soils is restricted by other mineral nutrient deficiencies that include molybdenum, calcium, potassium and sulphur which generally limit plant growth and yield [44]. According to Nkaa *et al.* [45], edaphic factors like P, Ca, K and S play a very critical role in stimulating cowpea productivity by promoting *Rhizobium*-legume symbiosis and initiating nodule formation. As indicated by Karikari *et al.* [43] and Willey [46], all attempts to improve cowpea should also consider the genetic manipulation of the crop, and its growing environment, especially the physiochemical characteristics of the soil.

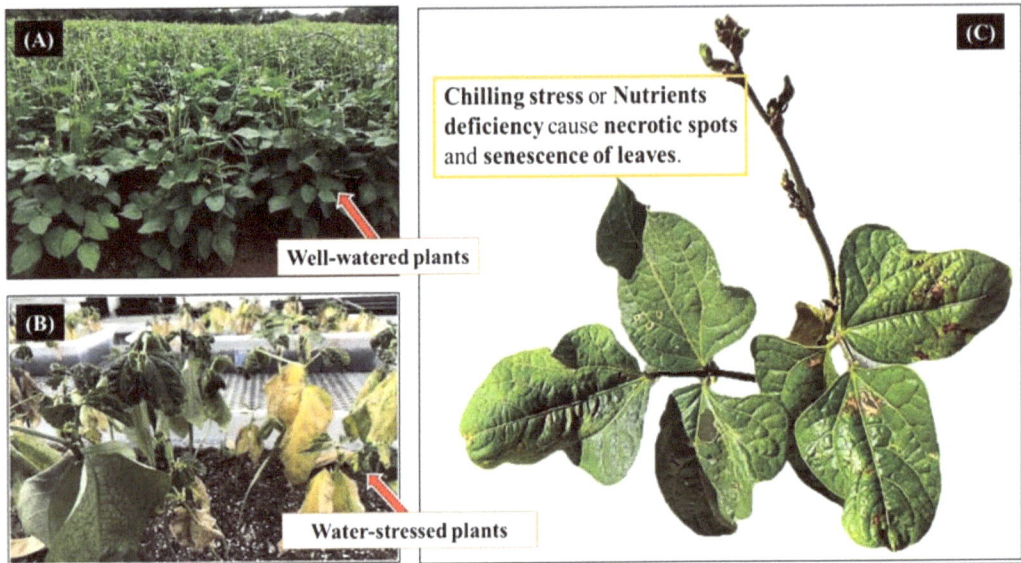

Fig. (3). Effect of abiotic stress (water deficit, chilling and nutrient deficiency) and well-irrigated soils during the growth of cowpea plants [3, 33].

COWPEA BREEDING

There are major opportunities for breeders to develop cowpea cultivars tolerant to various abiotic and biotic stress factors, as well as breeding for enhanced agronomic characteristics specifically adapted to different ecological and geographical zones [47]. Breeding methods such as mass selection, pure line breeding and pedigree selection are applied to self-pollinating cowpeas' genetic improvement [22]. Genetic breeding is conducted with the sole primary objective of achieving high grain yields, improved seed quality and stress resistance. Furthermore, the general strategy of most breeding programs is to develop various high yielding cowpea crops with the ability to adapt to different agro-ecological zones with regionally preferred traits for plant type, growth habits, and days to maturity and seed type [47]. Apart from the progress made in the breeding of improved varieties of cowpeas with multiple pests and disease resistance, cowpea is still subjected to considerable grain yield losses from several insect pests such as the cowpea pod borer, flower bud thrips and the pod sucking bug complex (Fig. **2**).

A high level of resistance to several insect pests exists in wild species, however, strong barriers also exist which prevent the transfer of these genes into cultivated cowpea [12, 22]. Opportunities introduced for biotechnology in cowpea improvement, especially in the development of suitable bioassays that helped with the identification of target traits for overcoming production constraints remain of great importance [48]. According to Ferry and Singh [49], sources of these targeted agronomic traits were identified through screening and selection of the germplasm materials available in different countries for use as breeding resources. Approximately 15,000 accessions of cultivated cowpea and more than 2000 wild relatives are maintained in various genetic resource banks, including the International Institute of Tropical Agriculture (IITA). These conserved genetic materials have made possible the identification of various plant genetic resources showing resistance to biotic and abiotic constraints [50, 51].

Consequently, the IITA together with the Bean/ Cowpea CRSP, advanced laboratories in the USA and Australia, African Agricultural Technology Foundation (AATF), Agricultural Research Council (South Africa), Network for Genetic Improvement of Cowpea for Africa (NGICA) and the Monsanto Corporation work independently or collaborate to utilize biotechnological strategies in complementing conventional methods for the genetic improvement against insect pests in cowpea [51]. The strategies focus more on pest control using synthetic genes, marker selections and bio-control agents as well as host plant resistance through the integration of traditional breeding methods and biotechnological applications [47, 51]. Nonetheless, as these private, parastatal and government-owned research institutions thrive to avert the negative effects of these yield-limiting constraints, many cowpea growers still reduce or prevent their impact through unsustainable and cost-ineffective use of agro-chemicals, irrigation systems, mechanisation, and the cultivation of genetically improved cultivars in areas where farmers are well-resourced.

Genomic Resources for Cowpea

The genomic resources of cowpea in comparison to other legume crops have been developed recently, with just a few currently available for use by researchers focusing on cowpea breeding. However, advances made to this point in cowpea genomics and biotechnology have efficiently and rapidly reduced the costs of evaluating cowpea populations as well as other legumes for genetic improvement.

Table **3** shows examples of some of the genomic resources developed from 2005 to 2015 to complement breeding approaches in cowpea and other crops [21]. These techniques and programs have already achieved noticeable progress in the fight against biotic stress diseases and plant attacks like ashy stem blight, cowpea bacterial blight (CoBB) and macrophomina. Additionally, Chamarthi *et al.* [53], reported successful breeding for control of parasitic weeds (Alectra and Stringa) and insect pests such as aphids, leaf thrips as well as flower thrips.

CoBB is a plant disease caused by *Xanthomonas axonopolis*, whilst ashy stem blight is instigated by *Macrophomina phaseolina*. Although these disease agents usually show pathogenic specificity, they may also cause severe damage, tissue senescence and subsequent deaths of plants, but mainly in selected crop genotypes. However, numerous reports, such as those of Boukar *et al.* [51], Chamarthi *et al.* [52] and Agbicodo *et al.* [53] asserted the use of marker-assisted selection (MAS) to identify biotic stress resistance loci to the pathogens and highlighted the development of QTLs that map identified DNA sections with the preferred phenotypes. Agbicodo *et al.* [53] screened eleven cowpea genotypes for resistance to CoBB using *Xav18* and *Xav19* virulent strains of *X. axonopolis* isolated from Kano (Nigeria), and identified cultivar IT81D-122-14, Aloka local and Danila as the most disease-resistant genotypes. CoBB remains one of the most important diseases in cowpea production, remaining prevalent in many parts of the world. This disease has caused more than 65% yield production losses in West Africa alone [54].

Naturally occurring resistance alleles could be mapped for exploitation in plant breeding and introgression into elite cultivated genotypes to improve agricultural performance. Many breeders seek to incorporate a wide range of abiotic and biotic stress resistance characters. A single feature polymorphism (SFP) in cowpea using a readily available soybean genome array was reported by Das *et al.* [55]. This study identified approximately 1000 SFPs for cowpea using this microarray-based marker to accelerate map-based cloning. These high-density genetic linkage maps are required for marker-assisted breeding, particularly to provide breeders with the ability to analyse gene inheritance and progeny-parent genome linkages. Further studies using methods demonstrated in Table **3** are needed to study the genetic diversity in cowpea, establish genomic libraries and characterise population lines based on their genomic variations in stress resistance/tolerance.

Biotechnological Approach

Most cowpea breeders employ backcross and multiple generations of ancestors (pedigree) to increase the certainty of pure lines and accelerate the expression of desirable traits in the progenies during breeding. As indicated earlier, these improvement efforts are continuously investigated and implemented due to the production constraints resulting in growth and yield inhibitions experienced by cowpea and other crops. However, several biotechnological strategies have also been developed to efficiently increase cowpea enhancement and productivity. The introgression of novel gene combinations using genetic engineering or recombinant DNA technology that allows the use of specific genes normally transformed in binary plasmid vectors of bacterial origin is tested. This technology involving the identification, isolation, reconstruction and transfection of exogenous DNA segments into hosts is also carried out by many laboratories worldwide, in the attempt to advance plant tolerance to stress factors [56].

Table 3. Cowpea genomic resources that were released between 2005 and 2015, as reported by Boukar *et al.* **[12].**

Resources	Short Description	Uses
Cowpea Genespace/Genomics Knowledge Base (CGKB)	Genetic markers, gene-space, metabolic pathways, mitochondrial and chloroplast sequences.	Tool for gene discovery, enzyme and metabolic pathway.
Cowpea consensus genetic linkage map	A consensus map containing 1107 EST-derived SNP markers on 11 linkage groups was constructed from 13 population-specific maps	For QTL identification, map-based cloning, diversity and association mapping.
HarvEST:Cowpea	EST database with gene function analysis and primer design	Online cowpea genomics browser.
Microarray chip	41,949 EST sequences from drought-stressed and non-stressed drought susceptible and tolerant cowpea materials generated, representing 16,954 unigenes.	For expression analysis in cowpea
Physical map of cowpea	Fingerprinted physical map of 60,000 BACs from IT97K-499-35, with the final assembly map of 43,717 BACs with a depth of 11x genome coverage	Tool for gene discovery.
The Cowpea Genomics Initiative (CGI)	Some advances in cowpea genomics.	Tools for gene discovery and cowpea improvement.
Software Programme	SNP Selector, 'KBioConverter', and 'Backcross Selector' used for the management of genotyping data.	For molecular breeding.
Validated SSR marker kit	Reference kit of 20 SSRs used to define the Cowpea Germplasm Reference Set representing the genetic diversity of the entirety of the IITA cowpea germplasm bank collection.	For diversity analysis and gene discovery

Genetic transformation holds the great potential to be systematically applied and optimized for the development of new agronomically useful cowpea varieties. Attempts to produce new breeding materials in legumes using a variety of transformation techniques have been pursued. But, most of these crops have proved to be highly recalcitrant or resistant to genetic transformation. This is so because the increases in legume yields still rely on chemical applications (fertilisers, herbicides and pesticides) which generate various economic and ecological problems, particularly the problem of environmental pollution. Among the techniques used, *Agrobacterium tumefaciens*, a Gram-negative bacterium, offers promise for efficient delivery of desired genes into some legume plants' genomes. *Agrobacterium*-mediated genetic transformation attempts can be carried out under *in vitro* and *in vivo* culture conditions. As for conventional breeding, improvements through *in vivo* and *in vitro* methods will overcome breeding limitations and lead to genetic modifications conferring tolerance to environmental stress, like drought resistance, pests and improving seed quality and overall yields.

A novel biolistic-mediated genetic improvement system in cowpea was reported by Ivo *et al.* [57]. The system combined the use of herbicide imazapyr to select transgenic meristematic cells after biolistic introgression of mutated *ahas* gene coding for acetohydroxyacid synthase under the control of *ahas* '5 regulatory sequence. Biotic stress-resistant features adapted from several legumes and transformation systems were reproducibly used to produce transgenic cowpeas. These transgenic cowpeas contained the *bar* gene and expressed phosphinothricin acetyl transferase (PAT) enzyme activity that confers resistance towards phosphinothricin and bialaphos with the normal production of tropane and alkaloids [58]. The biological compounds like alkaloids are commonly used as pest control agents for insect pests, plant diseases and weeds. Bett *et al.* [59] reported an improved sonication-assisted *Agrobacterium*-mediated gene delivery using cotyledonary node explants derived from imbibed cowpea seeds. Transgenic cowpeas encoding an insecticidal protein from *Bacillus thuringiensis* were effectively regenerated and selected using geneticin that was alternated with kanamycin, as selective markers.

A much faster and easier DNA modifying technology that took the scientific community by storm is the Clustered Regularly Interspaced Short Palindromic Repeat (CRIPR) and flanked by CRIPR associated genes (Cas). CRIPR-Cas is a unique set of partially palindromic repeated DNA sequences found in the genome of bacteria and other microorganisms [60]. The RNA-guided CRIPR-Cas nuclease system is one of the genome editing methods that have emerged in the recent years to induce targeted DNA double-stranded breaks at specific genomic loci. Among these, the small RNAs guided Cas9 nucleases serve as a high-throughput, efficient and highly specific gene-editing technique for a variety of cell types and organisms using a Watson-Crick base pairing [61]. Ji *et al.* [62] demonstrated the possibility of CRISPR-Cas9 system application in cowpea genome editing for mutants generation by disrupting the representative symbiotic nitrogen fixation (SNF) genes. Nodule formation was completely blocked in the mutants by targeting both alleles of the symbiosis receptor-like kinase. The rapidly evolving sequence-specific nucleases based genome editing and optimisation of other genetic manipulation technologies will certainly make insect-virus resistance genes available and compatible for the development of newly improved cowpea genotypes.

SUMMARY AND FUTURE PROSPECTS

Although significance progress has been achieved in the last three decades in the development of new *Vigna unguiculata* cultivars, refinement and establishment of highly efficient genetic manipulation technologies are still outstanding. Cowpea represents a crucial source of proteins, fibre and carbohydrates for many African populations. Thus, both conventional and modern breeding of cowpea must bring about noticeable production improvements. Cowpeas with increased yield potential to feed the growing populations remains a necessity. These genetically improved plants should ideally exhibit resilience to climate change-induced stresses, must grow on nutrient inadequate soil and show resistance to insect pests as well as viral or bacterial diseases (Fig. **4**). Shunmugan *et al.* [63] emphasised that the physiological breeding strategies that were proven to be successful in cereal transformation need to be adapted to underutilised food legumes such as cowpea improvement to realise their potential, especially in a collaborative approach between plant breeders and plant physiologists.

Molecular breeding (QTLs) to map the relevant physiological traits and the potential of CRISPR-Cas9 gene editing could be effectively used to achieve sustainable environmental and nutritional food security in all legumes. Cowpea is a crop that has wide environmental adaptability, particularly high tolerance to areas with extreme heat and drought [62, 63]. This means that it could serve as one of the most important genetic resources and reservoir of genes of interest for abiotic stress improvement of many recalcitrant legume crops like soybean. This crop can also be used as a donor source of exogenous DNA fragments that may be introduced in other legume crops to confer heat and drought tolerance, including some moderately abiotic stress susceptible cereals. But, future progress will also depend upon the development and sustainability of multi-disciplinary collaborations between researchers, farmers and consumers, as well as the integrated scientific research in fields including genetics, plant breeding, engineering, biometrics and bioinformatics.

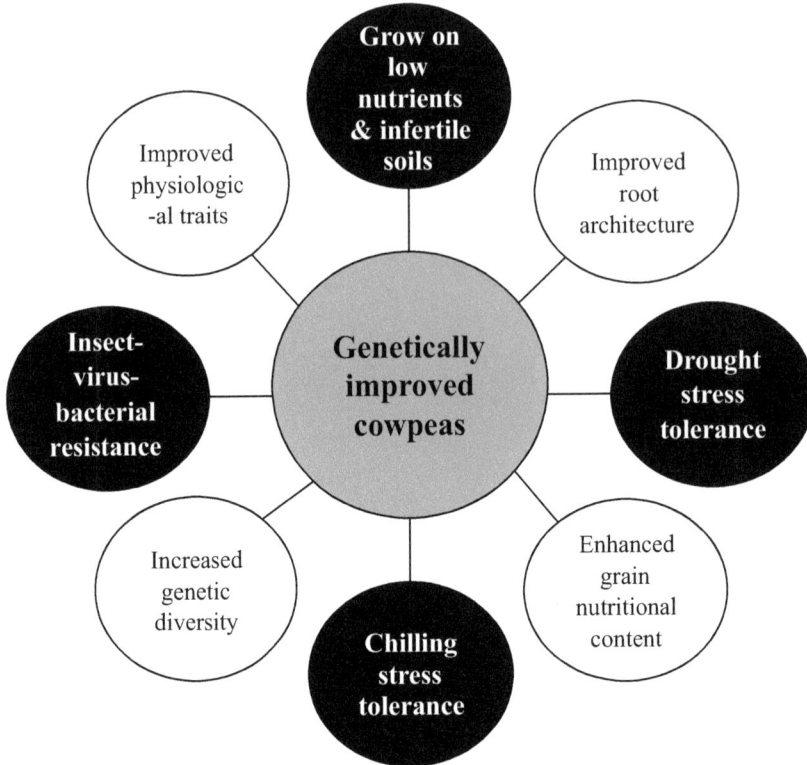

Fig. (4). Illustration of envisioned high yielding genetically improved cowpea plants showing resilience to harsh climatic conditions, bacteria/virus-induced diseases and insect attacks.

New forms of communication and professional collaborations must be established so that, genetic engineering, genome editing and marker-assisted breeding, *etc.* can achieve their respective potential. Unlike abiotic stress, direct control of biotic stress factors such as insects and virus is impossible due to the fact that virucidal chemicals are not yet available [64] and different insects may adapt, while they attack plants at different stages of their life cycle, making pesticide application difficult and costly. Therefore, improved cowpea growth and yield depend on a myriad of factors and not limited to genotype performance and planting conditions. This also suggests that research on underutilised crops like cowpea lags far behind compared to other legumes with sequenced genomes like *Glycine max*, *Medicago truncatula* and *Lotus japonicus* [62]. Furthermore, with more information on the genetic sequencing of cowpea (HarvEST:Cowpea, Table **3**) currently available from combined genetic resources, the development of abiotic stress-resilient mutants may be attained and used as a model species for genomic and breeding research. Such achievements could be extended to the breeding of cowpea for biotic stress resistance.

LIST OF ABBREVIATIONS

AATF	African Agriculture Technology Foundation
C$_3$ plants	Three carbon (3-phosphoglycerate) metabolism
CAM	Crassulacean acid metabolism
CoBB	Cowpea bacterial blight
CRISPR-Cas9	Clustered regularly interspaced short palindromic repeats-CRISPR associated protein 9
FAO	Food and Agriculture Organisation
IITA	International Institute of Tropical Agriculture
MAS	Marker-assisted selection
NGICA	Network for genetic improvement of cowpea for Africa
PAT	Phosphinothricin acetyl transferase
QTL	Quantitative trait locus
SSA	Sub-Saharan Africa

CONSENT FOR PUBLICATION

Not applicable.

CONFLICT OF INTEREST

The author declares no conflict of interest, financial or otherwise.

ACKNOWLEDGEMENTS

Declared none.

REFERENCES

[1]   Taiz L, Zeiger E. Plant physiology. 6th ed. Sunderland, Massachusetts, USA: Sinnauer Associations Inc. 2014; pp. 459-81.

[2]   Kumar J, van Rheenen HA. A major gene for time of flowering in chickpea. J Hered 2000; 91(1): 67-8.
[http://dx.doi.org/10.1093/jhered/91.1.67] [PMID: 10739130]

[3]   Boukar O, Belko N, Chamarthi S, *et al.* Cowpea (*Vigna unguiculata*): Genetics, genomics and breeding. Plant Breed 2018; 1-10.

[4]   Mwale SE, Shimelis H, Mafonoya P, Mashilo J. Breeding tepary bean (*Phaseolus acutifolius*) for drought adaptation: A review. Plant Breed 2020; 00: 1-13.
[http://dx.doi.org/10.1111/pbr.12806]

[5]   Nilsen E, Orcutt D. Physiology of plants under stress: Abiotic factors. Wallingford, UK: CABI Publication 1996; p. 689.

[6]   Hall AE, Cisse N, Thiaw S, Mcwatters KH. Development of cowpea cultivars and germplasm by the cowpea CRSP. Field Crops Res 2003; 82: 103-34.

[http://dx.doi.org/10.1016/S0378-4290(03)00033-9]

[7]   Quin FM. Introduction. In: Singh BB, Mohan-Raj DR, Dashiell KE, Jackai LEN, Eds. Advances in cowpea research Co-publication of International Institute of Tropical Agriculture (UTA) and Japan International Research Centre for Agricultural Sciences (JIRCAS). Ibadan, Nigeria: UTA 1997; pp. ix-xv.

[8]   OECD. Common Bean (Phaseolus vulgaris): Safety assessment of transgenic organisms in the environment. OECD Consensus Document 2015; Vol. 6.

[9]   Celmeli T, Sari H, Canci H, *et al.* The nutritional content of common bean (*Phaseolus vulgaris* L.) landraces in comparison to modern varieties. Agron 2018; 8(9): 166.
        [http://dx.doi.org/10.3390/agronomy8090166]

[10]   Awonaike KO, Kumarasinghe KS, Danso SKA. Nitrogen fixation and yield of cowpea (*Vigna unguiculata*) as influenced by cultivar and *Bradyrhizobium* strain. Field Crops Res 1990; 24(3-4): 163-73.
        [http://dx.doi.org/10.1016/0378-4290(90)90035-A]

[11]   Karoney EM, Ochieno DMW, Baraza DL, Muge EK, Nyaboga EN, Nalanyange V. Rhizobium improves nutritive sustainability and tolerance of *Phaseolus vulgaris* to *Colletotrichum lindemuthianum* by boosting organic nitrogen content. Appl Soil Ecol 2020; 149: 103534.
        [http://dx.doi.org/10.1016/j.apsoil.2020.103534]

[12]   Food and Agriculture Organization of the United Nations (FAO). Cowpea: Post-harvest operations. Rome, Italy: AGST/FAO 2004; pp. 1-71.

[13]   Department of Agriculture, Forestry and Fisheries (DAFF). Production guidance for cowpeas. Directorate Agricultural Information Services, Republic of South Africa, Pretoria 2019.

[14]   Langyintuo AS, Lowenberg-DeBoer J, Faye M, *et al.* Cowpea supply and demand in West and Central Africa. Field Crops Res 2003; 82: 215-31.
        [http://dx.doi.org/10.1016/S0378-4290(03)00039-X]

[15]   Carsky RJ. Response of cowpea and soybean to P and K on *terre de barre* soils in southern Benin. Agric Ecosyst Environ 2003; 100: 241-9.
        [http://dx.doi.org/10.1016/S0167-8809(03)00192-0]

[16]   Sanou J, Bationo BA, Barry S, Nabie LD, Bayala J, Zougmore R. Combining soil fertilization, cropping systems and improved varieties to minimize climate risks on farming productivity in northern region of Burkina Faso. Agric & Food Secur 201 5(20): 1-12.
        [http://dx.doi.org/10.1186/s40066-016-0067-3]

[17]   Saka JO, Agbeleye AO, Ayoola OT, Lawal BO, Adetumbi JA, Oloyede-Kamiyo QO. Assessment of varietal diversity and production system of cowpea (*Vigna unguiculata* (L.) Walp.) in southwest Nigeria. J Agric Rural Dev Trop Subtrop 2018; 119(2): 43-52.

[18]   Namugwanya M, Tenywa JS, Otabbong E, Mubiru DN, Basamba TA. Development of common bean (*Phaseolus vulgaris* L.) production under low soil phosphorus and drought in Sub-Saharan Africa: A Review. J Sustain Dev 2014; 7(5): 128-39.
        [http://dx.doi.org/10.5539/jsd.v7n5p128]

[19]   Fatokun CA, Menancion DI, Danesh D, Young ND. Evidence for orthologous seed weight genes in cowpea and mung bean based on RFLP mapping genetics 1993; 132(3): 841-6.

[20]   Abate T, Alene AD, Bergvinson D, Shiferaw B, Orr A, Aasfaw S. Tropical grain legumes in Africa and South Asia: Knowledge and opportunities Research Report. Nairobi: ICRISAT 2012.

[21]   Boukar O, Fatokun CA, Huynh B-L, Roberts PA, Close TJ. Genomic tools in cowpea breeding programs: Status and perspectives. Front Plant Sci 2016; 7(757): 757.
        [http://dx.doi.org/10.3389/fpls.2016.00757] [PMID: 27375632]

[22]   Boukar O, Fatokun C. Strategies in cowpea breeding In: Zerihum T, Ed. Proceedings of an

international conference. Bern, Switzerland. 2007.

[23]   Jackai LEN, Daoust RA. Insect pests of cowpeas. Annu Rev Entomol 1986; 31: 95-119.
       [http://dx.doi.org/10.1146/annurev.en.31.010186.000523]

[24]   Singh Sk, Nene YL, Reddy MV. Influence of cropping system on *Macrophomina phaseolina*
       populations in soil. Plant Dis 1990; 74(814): 1-3.
       [http://dx.doi.org/10.1094/PD-74-0812]

[25]   Singh SR, Van Emden HF. Insect pests of grain legumes. Annu Rev Entomol 1979; 24: 255-78.
       [http://dx.doi.org/10.1146/annurev.en.24.010179.001351]

[26]   Singh SR, Allen DJ. Cowpea pests and diseases: Manual series no 2 International Institute of Tropical
       Agriculture. Ibadan, Nigeria: IITA 1979; p. 113.

[27]   Harveson B. Cowpea bacterial wilt- An old disease in a new crop. Panhandle Research and Extension
       Centre, University of Nebraska-Lincoln 2018.

[28]   Fink W, Haug M, Deising H, Mendgen K. Early defence responses of cowpea (*Vigna sinensis* L.)
       induced by non-pathogenic rust fungi. Planta 1991; 185(2): 246-54.
       [http://dx.doi.org/10.1007/BF00194067] [PMID: 24186348]

[29]   Mohammed A. An overview of distribution, biology and the management of common bean
       Anhracnose. J Plant Pathol Microbiol 2013; 4(8): 1-6.
       [http://dx.doi.org/10.4172/2157-7471.1000193]

[30]   Nsa IY, Kareem KT. Additive interactions of unrelated viruses in mixed infections of cowpea (*Vigna
       unguiculata* L. Walp). Front Plant Sci 2015; 6(812): 812.
       [http://dx.doi.org/10.3389/fpls.2015.00812] [PMID: 26483824]

[31]   Osdaghi E, Young AJ, Harveson RM. Bacterial wilt of dry beans caused by *Curtobacterium
       flaccumfaciens* pv. *flaccumfaciens*: A new threat from an old enemy. Mol Plant Pathol 2020; 21(5):
       605-21.
       [http://dx.doi.org/10.1111/mpp.12926] [PMID: 32097989]

[32]   Togola A, Boukar O, Belko N, *et al.* Host plant resistance to insect pests of cowpea (*Vigna
       unguiculata* L. Walp.): Achevements and future prospects. Euphytica 2017; 213(239): 1-16.
       [http://dx.doi.org/10.1007/s10681-017-2030-1]

[33]   Revelombola W, Shi A. Investigation on various aboveground traits to identify drought tolerance in
       cowpea seedlings. HortScience 2018; 54(12): 1757-65.
       [http://dx.doi.org/10.21273/HORTSCI13278-18]

[34]   Chikutuma M, Tembo L, Karangwa W. The effect of herbicides on residual effects of atrazine under
       conservation agriculture. Greener J Agric Sci 2015; 5(2): 62-75.

[35]   Khan HR, Paull JG, Siddique KHM, Stoddard FL. Faba bean breeding for drought affected
       environments: A physiological and agronomic perspective. Field Crops Res 2010; 115: 279-86.
       [http://dx.doi.org/10.1016/j.fcr.2009.09.003]

[36]   Ashraf MA. Waterlogging stress in plants- A review. Afr J Agric Res 2012; 7: 1976-81.

[37]   Grahama PH, Rosasb JC, Estevez de Jensen C, *et al.* Addressing edaphic constraints to bean
       production: The bean/ cowpea CRSP project in perspective. Field Crops Res 2003; 82: 179-92.
       [http://dx.doi.org/10.1016/S0378-4290(03)00037-6]

[38]   Bastos EA, Nascimento SP, Silva EM, Freire Filho FR, Gomide RL. Identification of cowpea
       genotypes for drought tolerance. Cienc Agron 2011; 42: 100-7.
       [http://dx.doi.org/10.1590/S1806-66902011000100013]

[39]   Nascimento SP, Bastos EA, Araújo ECE, Freire Filho FR, da Silva EM. Tolerance to water deficit of
       cowpea genotypes. Rev Bras Eng Agric Ambient 2011; 15(8): 853-60.
       [http://dx.doi.org/10.1590/S1415-43662011000800013]

[40] Anyia AO, Herzog H. Water-use efficiency, leaf area and leaf gas exchange of cowpeas under mid-season drought. Eur J Agron 2004; 20: 327-39.
[http://dx.doi.org/10.1016/S1161-0301(03)00038-8]

[41] Hall AE. Cowpea. In: Smith DL, Hamel C, Eds. Crop yield. Berlin, Heidelberg: Springer 1999; pp. 355-73.
[http://dx.doi.org/10.1007/978-3-642-58554-8_12]

[42] Warrag MOA, Hall AE. Reproductive responses of cowpea (*Vigna unguiculata* (L.) Walp.) to heat stress. II. Responses to night air temperature. Field Crops Res 1984; 8: 17-33.
[http://dx.doi.org/10.1016/0378-4290(84)90049-2]

[43] Karkari B, Arkorful E, Addy S. Growth, nodulation and yield response of cowpea to phosphorus fertiliser application in Ghana. J Agron 2015; 14(4): 234-40.
[http://dx.doi.org/10.3923/ja.2015.234.240]

[44] Sanginga N, Lyasse O, Singh BB. Phosphorus use efficiency and nitrogen balance of cowpea breeding lines in a low P soil of the derived Savanna zone in West Afric. Plant Soil 2000; 220: 119-28.
[http://dx.doi.org/10.1023/A:1004785720047]

[45] Nkaa FA, Kumar RAJ, Kalloo G. Diet versatility in cowpea (*Vigna unguiculata*) genotypes. Indian J Agric Sci 2001; 71: 598-601.

[46] Willey RW. Intercropping- its importance and research needs- competition and yield advantages. Field Crop 1979; 32: 1-10.

[47] Timko MP, Singh BB. Cowpea, a multifunctional legume. In: Moore PH, Ming R, Eds. Genomics of tropical crop plants. Cham: Springer 2008; pp. 227-58.
[http://dx.doi.org/10.1007/978-0-387-71219-2_10]

[48] Gatehouse JA. Biotechnological prospects for engineering insect-resistant plants. Plant Physiol 2008; 146(3): 881-7.
[http://dx.doi.org/10.1104/pp.107.111096] [PMID: 18316644]

[49] Ferry RL, Singh BB. Cowpea genetics: A review of the recent literature. In: Singh BB, Mohan-Raj DR, Dashiell KE, Jackai LEN, Eds. Advances in cowpea research Co-publication of International Institute of Tropical Agriculture (UTA) and Japan International Research Centre for Agricultural Sciences (JIRCAS). Ibadan, Nigeria: UTA 1997; pp. 13-29.

[50] Singh BB, Chambliss OL, Sharma B. Cowpea genetics: A review of the recent literature. In: Singh BB, Mohan-Raj DR, Dashiell KE, Jackai LEN, Eds. Advances in cowpea research Co-publication of International Institute of Tropical Agriculture (UTA) and Japan International Research Centre for Agricultural Sciences (JIRCAS). Ibadan, Nigeria: UTA 1997; pp. 30-44.

[51] Boukar O, Fatokun CA, Roberts PA, *et al.* Grain legumes, series handbook of plant breeding. New York: Springer-Verlag 2015; pp. 219-50.

[52] Chamarthi SK, Belko N, Togola A, Fatokun CA, Boukar O. Genomics-assisted breeding for drought tolerance in cowpea In: V Rajpal, D Sehgal, A Kuma, S Raina, Eds. Genomics assisted breeding of crops for abiotic stress tolerance, Vol 11, Sustainable development and biodiversity, Vol 21 . Springer 2019; pp. 187-209.

[53] Agbicodo EM, Fatokun CA, Bandyopadhyay R, *et al.* Identification of markers associated with bacterial blight resistance loci in cowpea. Euphytica 2010; 175: 215-26. [*Vigna unguiculata* (L.) Walp.].
[http://dx.doi.org/10.1007/s10681-010-0164-5]

[54] Sikirou R, Wydra K. Persistence of *Xanthomonas axonopodis* pv. *vignicola* in weeds and crop debris and identification of *Sphenostylis stenocarpa* as a potential new host. Eur J Plant Pathol 2004; 110: 939-47.
[http://dx.doi.org/10.1007/s10658-004-8949-9]

[55] Das S, Bhat PR, Sudhakar C, *et al.* Detection and validation of single feature polymorphisms in cowpea (*Vigna unguiculata* L. Walp) using a soybean genome array. BMC Genomics 2008; 9(107): 107.
[http://dx.doi.org/10.1186/1471-2164-9-107] [PMID: 18307807]

[56] Mangena P. A simplified *in-planta* genetic transformation in soybean. Res J Biotechnol 2019; 14(9): 117-25.

[57] Ivo NL, Nascimento CP, Vieira LS, Campos FAP, Aragão FJL. Biolistic-mediated genetic transformation of cowpea (*Vigna unguiculata*) and stable Mendelian inheritance of transgenes. Plant Cell Rep 2008; 27(9): 1475-83.
[http://dx.doi.org/10.1007/s00299-008-0573-2] [PMID: 18587583]

[58] Popelka JC, Gollasch S, Moore A, Molvig L, Higgins TJV. Genetic transformation of cowpea (*Vigna unguiculata* L.) and stable transmission of the transgenes to progeny. Plant Cell Rep 2006; 25(4): 304-12.
[http://dx.doi.org/10.1007/s00299-005-0053-x] [PMID: 16244884]

[59] Bett B, Gollasch S, Moore A, Harding R, Higgins TJV. An improved transformation system for cowpea (*Vigna unguiculata* L. Walp.) *via* sonication and a kanamycin-geneticin selection regime. Front Plant Sci 2019; 10(219): 219.
[http://dx.doi.org/10.3389/fpls.2019.00219] [PMID: 30873198]

[60] Barrangou R, Marraffini LA. CRISPR-Cas systems: Prokaryotes upgrade to adaptive immunity. Mol Cell 2014; 54(2): 234-44.
[http://dx.doi.org/10.1016/j.molcel.2014.03.011] [PMID: 24766887]

[61] Ran FA, Hsu PD, Wright J, Agarwala V, Scott DA, Zhang F. Genome engineering using the CRISPR-Cas9 system. Nat Protoc 2013; 8(11): 2281-308.
[http://dx.doi.org/10.1038/nprot.2013.143] [PMID: 24157548]

[62] Ji J, Zhang C, Sun Z, Wang L, Duanmu D, Fan Q. Genome editing in cowpea *Vigna unguiculata* using CRISPR-Cas9. Int J Mol Sci 2019; 20(10): 1-13.
[http://dx.doi.org/10.3390/ijms20102471] [PMID: 31109137]

[63] Shunmugam ASK, Kannan U, Jiang Y, Daba KA, Gorim LY. Physiology based approaches for breeding of next generation food legumes. Plants (Basel) 2018; 7(3): 1-22.
[http://dx.doi.org/10.3390/plants7030072] [PMID: 30205575]

[64] Ibrahim AB, Abdu SL, Usman IS, Aragao FJL. Horizontal gene transfer in cowpea (*Vigna unguiculata* L. Walp) through genetic transformation. Afr J Bot 2017; 5(5): 128-37.

CHAPTER 8

Transgenic Grain Legumes

Phetole Mangena[*] and **Esmerald Khomotso Michel Sehaole**

Department of Biodiversity, School of Molecular and Life Sciences, Faculty of Science and Agriculture, University of Limpopo, Private Bag X1106, Sovenga, 0727, South Africa

Abstract: Recombinant DNA technology remains one of the best tools that still presents a great potential to enhance genetic improvement in many recalcitrant crops since its discovery more than two decades ago. This chapter intends to provide a comprehensive review of the applications of genetic transformation techniques in legumes, useful for both growth and yield improvements, especially under abiotic and biotic stress conditions. This technology is very promising in mitigating the current and future challenges in agriculture, with proven records in the development of a number of cereal and legume crops, such as maize, sorghum, soybean, lentils, peas, chickpeas, common beans, and alfalfa. In the midst of all reported advantages, this technology is also faced with several concerns, criticism, and possible shortcomings emanating from its adoption and production of novel genetically modified cultivars, especially at farm and market levels. Issues such as the genetic integrity of the transgenic cultivars, undesirable mutations, biosafety, and moral beliefs regarding the production and consumption of GM crops are among the controversial topics faced by this biotechnological tool. In addition, a large number of genotypes still persist in being recalcitrant to genetic manipulations, pending a cost-effective, precise, highly-competent, and robust approach for the generation of fertile transgenic plants. However, this technology remains widely utilised in many countries despite the numerous speculative concerns raised by some scientists, health professionals, and environmentalists.

Keywords: *Agrobacterium*, Electroporation-mediated transformation, Genetically modified organisms, Particle bombardment, Transgenic crops.

INTRODUCTION

Legumes serve as the most important agricultural crops and the second largest group of domesticated crops after cereals (Poaceae), primarily for human consumption, manufacturing of livestock feeds, and as soil enhancing green manure. Legumes are agriculturally grown for both subsistence, commercial and proprietary farming throughout the world [1 - 3].

---
[*] **Corresponding author Phetole Mangena**: Department of Biodiversity, School of Molecular and Life Sciences, Faculty of Science and Agriculture, University of Limpopo, Private Bag X1106, Sovenga, 0727, South Africa; Tel: +2715-268-4715; E-mails: mangena.phetole@gmail.com & phetole.mangena@ul.ac.za

Legumes are usually cultivated under irrigation systems and chemically fertilised soils, particularly in the tropics and subtropical regions. The crops are easily sold in households, often serving as an income source for many families and small holder farmers. These plants exhibit tolerance to harsh and unfavourable environmental conditions, and can symbiotically interact with nitrogen-fixing bacteria, thus playing a vital role in maintaining soil nitrogen and fertility [4]. Legumes continually hold the potential to play a critical role in the eradication of malnutrition by decreasing the rising rates of poverty in developing countries.

This important role that these plants play is attributed to the high levels of proteins, minerals, fibre, carbohydrates, essential fatty acids, and vitamins contained within the seeds [2, 4]. Legumes serve as a cheaper, yet highly nourishing alternative for animal-based form of proteins and cereals to combat food insecurity. The lack of considerable amounts of essential nutrients in the body cause major nutritional conditions, such as protein-energy malnutrition (PEM), that affect many children in most African and Asian countries. According to Tran and Nguyen [1], PEM cases were reported to rapidly increase as a result of the rising costs of animal-based protein foods and cereal-based staples. These common foods are becoming highly unaffordable for low income households in poorer communities. Furthermore, legumes are also well-known for their high seed oil content mostly found in *Glycine max* L. (20–33%) and *Millettia pinnata* (25–35%). These legume species are preferably used as major sources of vegetable oil, and are both processed in the manufacturing of biofuel and other oil derivatives such as bio-lubricants [3, 5].

Based on the advantages outlined above, legume crops have been cultivated and commercialised at a rapid rate over the past three decades. Brookes and Barfoot [6] indicated that over 160 million hectares of land had been cultivated by 2014 which recorded a 94-fold increase compared to the previous years (1994–2010). But, the major breakthroughs in legume agriculture were achieved following the development and cultivation of genetically modified (GM) crops. Advances in agricultural biotechnology, like the establishment of genetic engineering tools used for the regeneration and production of transgenic crops (Table **1**), intensified the exploitation of legume crops. These improvements also brought about the various other molecular tools and techniques, which yielded crucial insights in the development of breeding materials used solely for selection and breeding of newly improved cultivars, production of important bio-stimulants/ biosynthetic compounds, as well as enhanced nutritional content and crop resistance against various environmental stress factors [7]. Currently, there is a significant number of GM varieties that are cultivated for both food and feed production in dicots and monocots as indicated in Table **1**.

Table 1. Some of the most common and widely cultivated monocots and dicots (GM) crops in developing countries since 2013.

Country	Area (Million Hectares)	GM Crops
Brazil	40.3	Soybean, maize, cotton
Argentina	24.4	Soybean, maize, cotton
India	11.0	Cotton
China	4.2	Cotton, papaya, poplar, tomato, sweet pepper
Paraguay	3.6	Soybean, maize, cotton
South Africa	2.9	Maize, soybean, cotton
Pakistan	2.8	Cotton
Uruguay	1.5	Soybean, maize
Bolivia	1.0	Soybean
Philippines	0.8	Maize
Burkina Faso	0.5	Cotton
Myanmar	0.3	Cotton
Mexico	0.1	Cotton, soybean
Colombia	0.1	Cotton, maize
Sudan	0.1	Cotton
Chile	<0.1	Maize, soybean, canola
Honduras	<0.1	Maize
Cuba	<0.1	Maize
Costa Rica	<0.1	Cotton, soybean
Romania	<0.1	Maize

Source: Azadi *et al.* [12].

According to ISAAA [8] and FAO [9], about 191.7 million hectares of world's arable land was cultivated with GM crops which were distributed in approximately 26 countries, and with pest resistant varieties being among the first introduced transgenic plants since 1996. These crops were considered as the fastest adopted crops in the history of modern agriculture. They presented a range of benefits, including the cultivation of pest-resistant varieties that have been grown to reduce both the application of synthetic insecticides and the production costs for farmers. Garcia-Yi *et al.* [10] emphasised that this rapid adoption of GM crops also resulted in substantial socio-economic impacts, generating a great deal of controversy. However, perceptions raised often appear to be more ideological than been based on scientific concerns, that are continuing to be echoed in the media and some academic press. This is due to the vast amount of published reports that are not reliable and evidence-based or give a true reflection of the socio-economic impacts of GM crops.

As GM crops hold the greatest potential to mitigate current and future challenges in agriculture, continued controversial publications in the media and press have already jeopardised the cultivation of GM crops, even leading to full and partial GMO ban in certain African and European countries [11]. Therefore, it is deemed necessary to use reliable and transparent methods of data collection as well as analysis to minimise potential bias in research or publications of controversial topics. Garcia-Yi *et al.* [10] reported the use of systematic maps employing structured procedures in the identification, selection and interrogation of collected evidence involved in studies dealing with controversial subjects such as the use and consumption of GMOs. These systematic procedures will effectively identify potential research gaps and efficiently guide future research. This chapter will, therefore, provide an important overview of the existing literature on the production, benefits and challenges facing the application of GM technology in grain legumes.

IMPORTANCE OF LEGUMES

Legumes are regarded as the second largest group of crops that are a major source of protein, vegetable oil, phytochemicals and nitrogen which contribute extensively to daily dietary requirements for human health. Tran and Nguyen [1] has indicated that in 2004, about 300 million mega grams of grain legumes were produced on about 13% of the cultivated land on earth accounting for about 27% of the world's primary crop production. The seeds of these cultivated legumes consist of all the essential minerals and secondary metabolites required by humans and animals in improving their wellbeing. They contain isoflavones reported to slow down tumour growth in both men and women, and the phytonutrients also mimic oestrogen effects, helping to strengthen the bones on women who reached post-menopausal period. In addition to their isoflavone content, soybean and other legumes are the source of all nine essential amino acids, needed for healthy muscles and bones. Maphosa and Jideani [4] indicated that the ability of legumes to prevent cancers, heart-related diseases and other degenerative diseases remain attributed to the biologically active phytochemicals that have demonstrated very high antioxidant properties, to counteract the effects of unstable reactive oxygen species (ROS).

Legume crops, as all vegetables, naturally contain zero-cholesterol, and this, along with their high fibre content, encourage bowel movements and reduces blood cholesterol levels elevated by consumption of foods with low-density lipoprotein (LDL) cholesterol. Furthermore, legumes provide hypoglycaemic effect reducing blood glucose levels which is beneficial to insulin-dependent diabetics [1, 13, 14]. Legumes flexibly blend and form part of a myriad of healthy eating plans. According to countless reports, such as those of Pathak [15], Maphosa and Jideani [4], these plants contain nutritional components essential for effective enzyme activity, iron metabolism, haemoglobin synthesis and oxygen transport throughout the body, and stabilisation of the plasma membrane. In addition to the many benefits, leguminous plants are generally low in saturated fats, with about 5–47% depending on the species.

According to Mabaleha and Yeboah [16], common bean (*Phaseolus vulgaris*) contains very low amounts of fats compared to all legume crops with only 1.6% (w/v) oil content on average. However, common bean plays a vital role in human nutrition as the other legumes based on other grain chemical compositions. Grain legumes with the highest oil content include soybean, lentils, lupins, and peas that have been potentially explored for biofuel production, in addition to alfalfa, which has served as a biomass resource to produce ethanol–biofuel through ligno cellulosic digestion [1, 4]. The production of GM legumes for commercial isation has benefited farmers around the world by generating higher yields, from land with improved weed management and at a lower costs. Christou [4] mentioned that the biotechnology of legume crops also aims to enhance yields by establishing crops that will no longer require the use of herbicides, pesticides, fertilizers, and antibiotics, which have harmful effects on the plant itself, consumers, and the environment.

Additionally, the crops produced will confer economic stability, storage viability even over prolonged periods and scientific sustainability [4, 6]. For subsistence or commercial farming of GM crops, economic gains are highly dependent on the specific traits of the crops generated and the good agricultural practises [10]. Some of the beneficial characteristics carried by the GM crops include herbicide tolerance, pest resistance, antibiotic resistance, and market-related abilities. However, the less privileged households in developing countries, which depend on legumes for survival, stand to benefit greatly from better yields and nutritionally enhanced GM crops. Not only will the income generated from the commercialisation of the GM crops increase for agribusiness or households, but the well-being of all who consume them will also be improved, particularly because food insecurity still remains one of the major current social concerns worldwide.

COMMON METHODS USED FOR LEGUME IMPROVEMENT

Numerous legume crops serve as important sources of high-quality proteins and oils for both humans and animals consumption and as active chemical ingredient formulations required for use in pharmaceutical and nutraceutical industries. For instance, soybean is a leguminous crop consisting of more than 36% proteins, 30% carbohydrates, and an excellent amount of dietary fibre, vitamins, and minerals. In addition, soybean also contains 20% oil content, which makes it the most important legume crop for the production of edible oil, with a great potential for use in the manufacturing of the much-needed eco-friendly biodiesel [17]. Attempts to produce new breeding materials in soybean and other legumes using a variety of transformation techniques have been pursued. However, most of these legume crops have proven to be highly recalcitrant or resistant to *in vitro* or *in vivo* genetic manipulations, particularly *Agrobacterium*-mediated genetic transformation. Fig. (**1**) gives a detailed schematic representation of soybean recalcitrance to *Agro bacterium*-mediated genetic transformation.

In general, seed viability decreases with the increase in storage duration of the seeds used to provide explants for genetic transformation from the day they were harvested. Additionally, the diagram shows that the capacity to form a callus and transgenic shoots from cotyledonary nodes transformed with *Agrobacterium* is also influenced by the components of the MS culture medium (antibiotics), explant type, *Agrobacterium*, and the genotype. Comparative protein analysis between infected and uninfected soybean coty-nodes also indicate varied protein profiles and significantly large amounts of proteins expressed according to genotypes and infection of explants with *Agrobacterium*. Thus, increasing soybean yield, like in any other legume crop, still depends on chemical applications that include fertilisers, herbicides, and pesticides, which typically generate various economic, health, and ecological problems, especially environmental pollution.

It is, therefore, crucial to enhance research in the genetic improvement of grain legumes in order to counteract recalcitrance to it, as well as to curb the adverse health and environmental effects caused as a result of the use of agro-chemicals. Genetic transformation offers great opportunities for rapidly introducing, selecting, or inducing desired characteristics in various leguminous plants for breeding purposes. The techniques allow for precise and controlled addition of the genes of interest to the genomes of targeted host plant species. Such genetically modified plants will possess new traits, including fungal resistance, nematode resistance, insect-pest resistance, and improved tolerance to the adverse environmental stress conditions. Amongst some of the abiotic environmental stresses, drought and salinity stress remain the most limiting factors in the growth and productivity of many legumes.

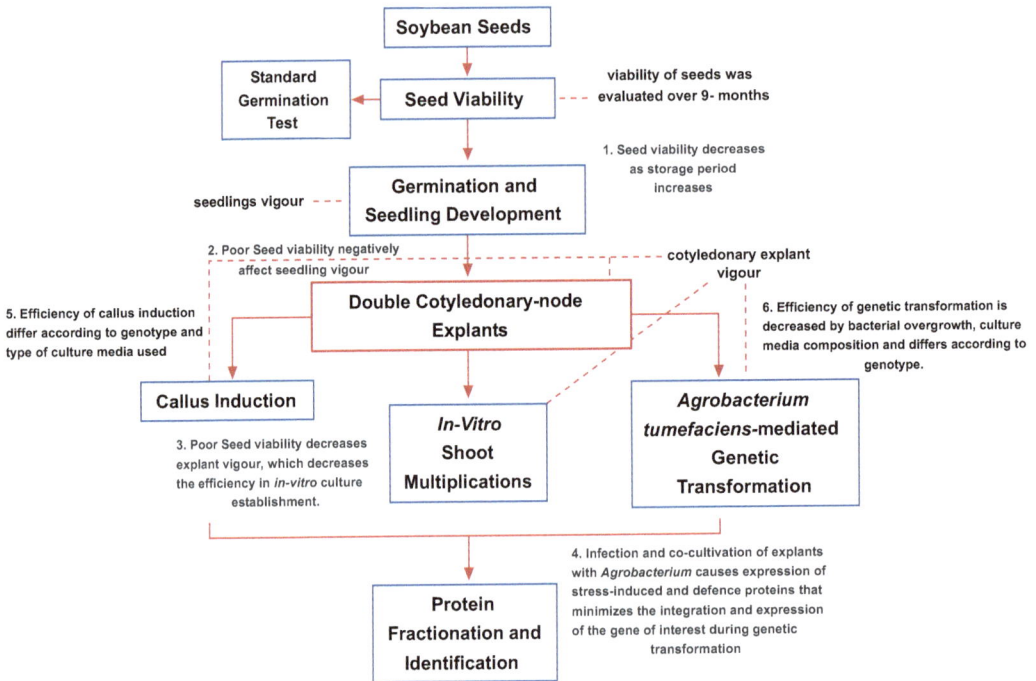

Fig. (1). Schematic representation of the assessment of the recalcitrance to genetic transformation in soybean.

Wieczorek and Wright [18] described genetic transformation as an efficient, quicker, and better approach to select and develop new plant breeding materials with better adaptation to diverse ecological conditions. Many different methods have been invented for the genetic transformation of forage and pulse legume crops. Microprojectile bombardment, electroporation and *Agro bacterium*-mediated genetic transformation are among the methods used for genetic manipulation of crops, including grain legumes. According to Finer and Dhillon [19] reports indicated that, successful transgene transfer and expression into host plants has been predominantly achieved mainly through the use of *Agrobacterium*- or particle bombardment- mediated genetic transformation. The other genetic transformation approaches are however, considered inefficacious, less widely used, highly expensive to carry out and inadequate for the efficient genetic transformation of many recalcitrant species [20]. Some of these most affordable and efficient protocols include *in-planta* transformation carried out by infiltrating plant tissues with *Agrobacterium* inoculum meanwhile, bypassing tissue culture conditions [21], and the floral dip technique reported by Zhang *et al.* [22] in the genetic transformation of *Arabidopsis thaliana* and *Medicago truncatula* using *Agrobacterium* suspension containing binary plasmid vector with the gene of interest for herbicide tolerance.

Agrobacterium-Mediated Genetic Transformation

Agrobacterium-mediated genetic transformation is a tool, developed in the early 1980s by Cheng *et al.* [23] to transfer vectors carrying specific regions of DNA for delivery into plant host cells (as exemplified in Fig. **1**). The method involves the use of two *Agrobacterium* species: *Agrobacterium tumefaciens* and *Agrobacterium rhizogenes*. The bacterial technique was first established by Hinchee *et al.* [24], reporting the first recovery of transgenic legume plants using cotyledonary node explants obtained from soybean cultivar Peking. Their experiment achieved about 6% of the recovered transgenic shoots using plant tissue culture. *Agrobacterium*-mediated genetic transformation serves as a method of choice for plant transformation because of the low copy number of DNA transferred, defined integration of transgenes, and its capacity to integrate foreign genes into the transcriptionally active chromosomal regions of the host plant's genome. Furthermore, the method is simple and affordable and has the potential to increase transformation efficiencies of many legumes.

This is so because it is easy to optimise using organic supplements such as L-cysteine, dithiothrietol, and acetosyringone, which help to reduce physiological stress in explant tissues and particularly, the effect of genotype specificity in many tested legume cultivars [22, 25, 26]. The production of stable transgenic legume plants continues to be a challenge as shown by the low rates or frequencies of genetic transformation obtained in legumes. Reports such as those of Paz *et al.* [20], Liu *et al.* [27], Untergasser *et al.* [28] and Tripathi *et al.* [29] stated that low recovery rates of transgenic plants are mostly attributed to poor explant response, genotype specificity and the inefficiencies in the regeneration protocols used. *Agrobacterium*-mediated genetic transformation using immature cotyledons [30] and axillary meristematic tissues [25, 31] as explants obtained from soybean seedlings have been the predominantly tested methods due to the explants' morphogenetic potentiality. Earlier research by Somers *et al.* [7], Paz *et al.* [20] on soybean transformation attempted to improve the efficiency of *in vitro* shoot regeneration and transformation rates using the abovementioned explant tissues. However, genetic transformation of soybean, like other legume crops, has always been recalcitrant and difficult to achieve due to factors mentioned earlier.

Electroporation-Mediated Transformation

Chowrira *et al.* [32] demonstrated electroporation-mediated transgene delivery in intact plant tissues that do not involve the use of plant tissue culture in pea, cowpea, lentil, and soybean. This approach is one of the most popular techniques used to deliver genes of interest from various sources into the cell's interior by disrupting membrane polarity using the electric field [33]. The delivery of foreign genetic materials is direct, and depends on the physical and chemical parameters to introduce biochemical substances like proteins, lipids as well as nucleic acids (RNA/DNA) into host's genome. Electro poration-mediated genetic trans formation has several advantages over other methods because there is no biological contact/interactions that are required, involves simple, quick, and highly efficient direct introgression of the transgene, and could cause membrane permeabilisation without potential cell complications. The cell membranes in plants are composed of phospholipids and amphipatic molecules that contain hydrophobic tails attached to the hydrophilic heads.

These membrane components can be able to be polarised when subjected to induced voltage of about 0.5 V under normal pressure and temperature, causing membrane permeability. Quecini *et al.* [34] reported the development of a highly efficient and reproducible genetic transformation for *Stylosanthes guianensis* using this direct physical method. They obtained more than 50% transformation efficiency assayed using GFP-coding gene *mgfp5* containing a CaMV35S constitutive promoter. This study also revealed the reliance of the protocol on energy input and electric field strength of 250 V.cm^{-1} discharged by 900–1000 µF capacitors to obtain the highest transient transformation frequencies.

Particle Bombardment

Biolistics, also known as particle bombardment or gene gun method entails the acceleration of high-density DNA coated carrier particles (of approximately two microns in diameter) passed through the cells, leaving the DNA inside bombarded tissues [33]. As reported by Rivera *et al.* [33], gene gun method was established in 1987 at the Cornell University for genetic engineering of single or organised cells in cereals. However, apart from the cereals, major crops like soybean and cotton plants have been among some of the most popular species that were genetically modified using this technique. According to Sanford [35], particle bombardment employs high velocity metal particles for delivery of active DNA segments into host plant tissues, which is a straightforward tissue engineering protocol. The procedure eliminates barriers involved in a number of genetic transformation systems, such as tissue culture-induced variations, genotype specificity, intensive laboratory protocols and time factors in the recovery of transgenic plantlets [2].

Rivera *et al.* [33] emphasised that, the comparison between *Agrobacterium* and particle bombardment in terms of transformation efficiency, transgene copy number, expression, inheritance, and physical structure of the transgenic loci using fluorescence *in situ* hybridisation shows that, in general, *Agrobacterium* offers significant advantages over biolistics. Nevertheless, organised, or single somatic cells, undifferentiated meristematic cells, embryos (from germ or somatic cells), callus cells and protoplasts could be used as targeted "explant" tissue for genetic transformation in both tools. As such, this approach can be successfully employed for both nuclear and chloroplast transformation [36]. Bhargava and Smigocki [37] reported a direct gene transfer of *GUS* and *NPTII* genes into *Vigna* species using germinated embryos. The report indicated that, factors such as a number of times in which embryo tissues were bombarded, gold particle-plasmid DNA ratio and pressure capacitor used to deliver coated DNA particles into tissues had an effect on the *GUS* expression frequency. In another study, Aragao *et al.* [38] clearly reviewed that, particle bombardment can be used to achieve genetic transformation of *Phaseolus vulgaris* L. with agronomically important traits from other legume varieties like when using *Agrobacterium*, electroporation, or polyethylene glycol (PEG)- mediated genetic transformation.

Other Novel Improvement Techniques and Recently Modified Traits

The production of GM crops has been a rapidly growing industry since its establishment and the first commercialised legume crop in 1996. Amongst the commercialised GM crops, many of the legumes were genetically improved to confer resistance against biotic stress (herbicide tolerance, pest resistance) and improved nutritional quality, including longer shelf life of GM foods [2, 4, 10]. Technologies such as biofortification, RNA interference (RNAi), gene editing, mutation breeding and omics have been established to play an imperative role in accelerating genetic modification of legumes and other crops [39]. These tools have been used to study plant protein synthesis pathways and understanding plant biological systems, whose in-depth knowledge has allowed farmers and plant biotechnologists to produce GM foods that benefit the economy at production, consumer, and environmental level [10, 40]. Although currently, techniques such as *Agrobacterium*-mediated genetic transformation apply a significant number of well-optimised and simplified steps/protocols in the production transgenic plants (Fig. **2**).

These protocols remain widely explored, where *in vivo*-based protocols (Fig. **2**) are efficiently and effectively used to bypass numerous challenges encountered during *in vitro* transformation (Fig. **1**). However, other techniques like biofortification involves processes that improves the nutrition content of food crops using either genetic engineering or classical breeding approaches, generating immeasurable improvements in vitamins and mineral nutritional status of transgenic crops. Biofortification differs from conventional fortification as the methods used focus on the genetic level for crop enhancement to express desired genes during growth and development [41]. For example, the biofortification of grain legumes and dryland cereals with iron and zinc was outlined and carried out by the International Crops Research Institute for the Semi-Arid Tropics (ICRISAT). Chickpea, soybean, pigeon pea and peas inoculated with nitrogen-fixing bacteria (*Pseudomonas sp., Brevibacterium sp., Bacillus sp., Enterobacter sp. and Acinetobacter sp.*) which colonised their roots and formed nodules to increase the seed mineral content (Fe and Zn) were reported by the ICRISAT [42].

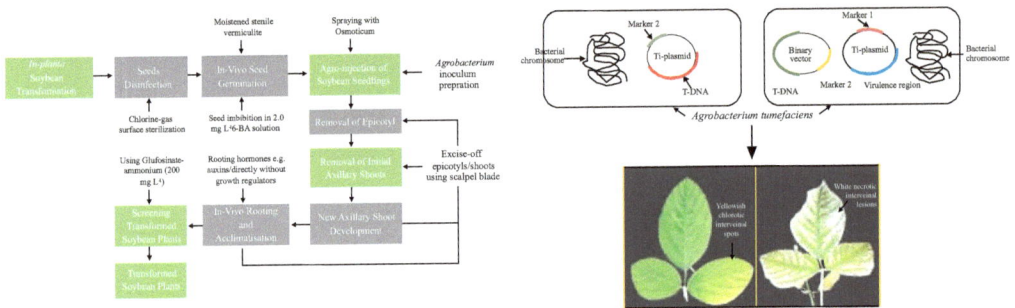

A.
Agro-injection of the T-DNA takes place through the axillary meristematic tissues found on the cotyledonary junctions of the seedlings. For seedling establishment, soybean seeds were imbibed overnight in a PGR solution.

B.
Foliar-application of 200 mg.L⁻¹ glufosinate-ammonium and responses of normal soybean leaf from transformed plants as well as the bleached leaf from control soybean plant showing irreversible lessions and death following glufosinate treatment. Schematic overview of the two vector system modifications in *Agrobacterium* stains for use in the transformation of major crops such as cotton, rice, maize, sugarcane, wheat and soybean, showing a binary vector containing the T-DNA region without virulence genes and a Ti-plasmid with a nonocogenic disarmed tumor-inducing plasmid containing the virulence genes.

Fig. (2). Example of a simplified *in planta* genetic transformation protocol (**A**), responses of soybean leaves to foliar-application of glufosinate-ammonium for *Agro*-injection and schematic representation of T-DNA transfer/expression (**B**).

Other crops include sorghum and cassava enhanced for amino acids and proteins, as well as the β-carotenoid-biofortification of sweet potato, and maize [41]. Apart from biofortification, the use of RNA interference (RNAi) has provided another avenue to control pests and diseases, by introducing novel plant traits and increase crop yield. RNAi is a biological process in which RNA molecules inhibit gene expression or translation by neutralizing targeted messenger RNA molecules. This is a posttranscriptional gene silencing technique that focuses on the incorporation of antisense RNA into host plant's genome in order to silence the expression of a single gene or a family of genes and down-regulate antinutrients, allergens as well as toxins [1, 39]. RNAi mediated silencing of the major allergens in soybean p34 protein was reported by Dunwell [43] and Newell-McGloughlin [40]. Inhibition of the expression of oleate desaturase that resulted in soybean plants with increased oleic acid content was also reported by Newell-McGloughlin [40].

Furthermore, the increased oleic acid content conferred natural resistance to oxidation and thermal degradation of proteins and other metabolites [40]. Tran and Nguyen [1] reported the down regulation of the expression of endogenous *FAD2-1* gene, which naturally functions to converts oleic acid into linoleic acid to increase oleic acid levels. All these reports suggest and indicate that the RNAi technology is capable of regulating protein synthesis, fatty acid metabolism, and modulating the expression of various endogenous genes for plant improvement. Expression of genes from unrelated organisms, *i.e.* bacteria and viruses, encoding desired traits, protein modification, and mutagenesis induced in target plants, are all used to introduce and express desirable characteristics in the progeny. These techniques involve molecular approaches to modify fatty acids, amino acids, and the genome to improve crops, including legumes in various specific ways.

INFLUENCE OF GM CROPS IN AGRICULTURE

Genetic transformation was primarily developed as a technology aimed at enhancing plant resistance to biotic and abiotic stress, accompanied by the need to acquire new desirable grain trait characteristics as well as high grain quality. The use of genetically modified organisms (GMOs) in agriculture took the field by a storm, generating many negative and positive opinions, while leading to some growers and consumers accepting them despite others remaining in opposition [12]. The major driving force in the continued cultivation of GM crops remains to be the increasing worldwide population, which continues to put pressure on agriculture that must meet the nutritional and health needs of the people. Since its introduction, this technology has improved commercial agriculture, food quality, and reliability by ensuring that food and nutrition security remain maintained, particularly in developing countries [44]. Furthermore, GM crops have the ability to increase species diversity by broadening the gene pool, which also has a simultaneous influence in reducing the negative environmental and ecological impacts [11].

Zdjelar *et al.* [45] emphasised that, poor species diversity in agriculture threatens the sustainability of the sector, leading to uniform genetic composition of crops, which promote crop's susceptibility to abiotic stress as well as the different pests and insect diseases. The African countries, especially South Africa, is among the developing countries that allow agricultural commercialisation of GM crops (Table 1). South African farmers continue to reap the benefits in high yields and quality of crop produce than their non-GM counterparts for crops like soybean, maize, and cotton [12]. Among the legumes, soybean has been a valuable source of carbohydrates, proteins, vitamins, oil *etc.*, and the most dominant GM crop by almost 50% cultivation on a global scale. Genetically modified legume crops collectively contribute to a more sustainable and diversified agriculture, especially during environmental stress caused by climate change. They remain prominent for efficiently providing healthy and affordable food resources in poorer communities.

The crops currently represent a better alternative crop production system that is cheaper, reliable, and more sustainable than conventional breeding systems [46]. Legumes assist in some of the far-reaching impacts in agriculture. Their traditional roles include (i) beneficial use in crop rotation systems, (ii) enhance yield stability of other crops, (iii) improve nutrient use efficiency, (iv) increase farm productivity by improving nutrient supply and physical characteristics of the soil, and (v) provide a stable ground cover which reduces the level of soil erosion by minimising water run off and rain drop impact [46, 47]. Legumes play numerous critical roles in agriculture and people's livelihoods due to their protein-rich seed content, and will definitely play an important role in the eradication of poverty, malnutrition, and nutrition-based children mortalities in developing countries.

Effects of GM Crops on Agri-Business and Economics

It has been over two decades since the introduction of GMOs in various countries, where both farmers and consumers found the technology very beneficial and significant. Garcia-Yi *et al.* [10] outlined the effects of GM crops at various subdivisions within the economy, which included the production, supply chain, consumer, and food security impacts. The report emphasised that the main impacts of GM crop production in the supply chain level originates from the costs of distribution and marketing, which in turn reflects a country's economic progress and potential. Within this level lies the main aspects with which the performance dynamics of GMO production are measured, namely (i) the efficiency of cost-effective delivery systems, (ii) the ability to provide quality products and (iii) the ability to reach consumer demands. According to economic research conducted from 1994 to 2014, the economic value of using GM crops accumulated during this period reached over US$150 billion [48].

Shukla *et al.* [48] reported that countries cultivating GMOs at a broader scale like India, have recorded an economic growth of about 15.8% between 2012 and 2013.

Among these, the application of genetically modified herbicide tolerant technology in soybeans has increased the income of many farmers by \$13.9 billion worldwide. The records indicated that 38% of this gross income was due to increases in yields and 62% was due to reduced costs in weed controls, including other farm management practices [6]. Reports also indicate that, herbicide tolerant (HT) GM cultivars increased farmer income by 47%; meanwhile 53% was due to cost effectiveness of the use of GM technology. Forster *et al* [50]. reported that farmers cultivating GM crops experienced more yield and profits than organic farmers whose production ranged between 20–25% on average. The report indicated that organic farmers obtain lower gross income compared to GM crop farmers, particularly due to poor crop growth because of biotic or environmental stress and yield.

In contrast to the GMO ban in some countries, South Africa like other developing countries (Table **1**) has regarded this as an exciting opportunity, which can be of great assistance in overcoming malnutrition and undernourishment. This greater economic benefits emanating from the use of biotechnology in agriculture has also drawn an interest from scientists to accelerate the production of genetically modified crops. In addition, this further trigger interest by scientists to genetically transform legumes for improved nutritional content and to generate phenotypically diverse plants, which are beneficial in studying gene expression and developing new genetic resources for breeding systems [7]. Thus, to maximise its socio-economic effects, genetic engineering together with other several application should remain investigated further for production purposes, consumer-related, environmental, and economic impacts. Conventional breeding programs, biofortification and other molecular techniques used to achieve genetic transformation have been largely successful, producing genetically enhanced crops that have improved growth, productivity, and nutritional content.

UPCOMING GMO TRENDS IN DEVELOPING COUNTRIES

With the nutritional and commercial value posed by leguminous crops coupled with increasing number of studies on plant biological systems and their function, the production of legumes continues to increase over time. This successful cultivation and commercialisation renders these crops economically vital and environmentally friendly [1, 3]. The economic importance, nutritional value, medicinal uses, cultural and physiological purposes played by crop legumes thus far is largely owed to their biological composition of bioactive compounds, whose roles in these platforms are unparalleled [1, 4]. Prior to the development of GM crops, legumes accounted for 27% of the basic global production with more than 300 million tonnes of grain yield [1]. After cereals, legumes have always formed part of the daily meals of various households worldwide, particularly in Afro-Asian and Latin American countries [4]. While some countries have adopted GM technology, some still depend on the production of natural legumes, for various reasons. For example, Zambia and Zimbabwe have banned the import of GMOs for ethical reasons, but in other countries, it is the lack of regulatory frameworks that will lead to comprehensive and balanced evaluations of GM products.

When GM technology was established, legumes formed part of the 160 million hectares of land domesticated for GM crops in 2012. One example of this is the production of glyphosate tolerant soybean crops where higher yields was obtained in reduced production time, recorded by the second generation of production [6]. Although HT soybean cultivation took place only later in some countries. This crop was produced in 2001 in countries like South Africa, as reported by the center for GMOs in South African Agriculture, and has significantly grown thus far. GMOs in South African Agriculture statistics previously projected that 90% of all cultivated soybeans would be GM crops by 2012/2013, which may considerably increase to 80% in 2020, reaching up to 2.9 million hectares of domesticated land (an increase of 21% from 2011, including maize and cotton). South Africa, along with Burkina Faso, Egypt, and Sudan constitute the few African countries producing and commercialising GM crops. While both Burkina Faso and Egypt had their first plantations in 2008, Sudan only had their first transgenic plantation in 2012 on 20 000 hectares of land. It indicated that the first GM crop planted in Egypt was maize on 700 hectares, increasing by 75% in 2012. Further considerations are in place and the targets are cassava and grapevines in several African countries, which could be a starting point for current non-GMO adopters. These data suggest that South Africa is the leading country in African GMO production, coming 9th after the global leader, United States.

POTENTIAL GMO APPLICATIONS AND RESTRICTIONS

Model legumes, such as *Medicago truncatula* and *Lotus japonicus,* have been useful in elucidating genetic engineering as they are autogamous in nature. These legume species regenerate in a relatively short period, have small diploid genomes, and exhibit a high volume of seed productivity [51, 52]. The plants continue to be readily responsive to genetic manipulation and offer the simplicity to utilise data obtained from their genetic expressions. They provide genomic insights required to understand the processes involved in the cloning of specific DNA segments and the introgression of complete gene segments into other legume species. Since the model plants are amenable to gene manipulation, their application to legume biotechnology studies has opened new avenues for the genetic manipulation of other crop legumes. Tran and Nguyen [1] reported that obtaining genomic data and understanding their potential might help reinforce genetic engineering as a tool for plant improvement and enable plant biotechnologists to gain deeper insights into genomic characteristics found in similar crops.

The overall application of biotechnology in improving legume plants has been quite critical, especially with the need to understand plant response mechanisms to be abiotic and biotic stress. This technology also plays a major role in agricultural sustainability, understanding plant biological systems and contributing to identifying abiotic stress or pathogen resistant genes [51]. Leguminous crops continue to be explored, manipulated, and enhanced for the biosynthesis of useful primary and secondary metabolites, following suitable and specific genomic modification objectives [53]. According to Song *et al.* [53], the production of seed oils can be increased in crops and miscellaneous legumes containing limited oil-seed storage bodies using biotechnology, and furthermore, employing the same approaches in introducing longer fatty acid chains in those plants. The genetic modification of oil fatty acid composition is now one of the widely explored technologies, whose main objective is to induce longer hydrocarbon chains in polyunsaturated fatty acids from staple crop legumes to manufacture biofuel.

Similarly, biotechnology can assist in the usage of oil bodies to transport vital proteins to the seed for yield quality improvement [53]. With these modifications, including other such as reduced LDL-cholesterol levels in GMO foods can be utilised for significant reduction in the risk of heart-related diseases in humans [4]. Experimental studies are ongoing by Monsanto, Pannar Seed, IITA and other private/government entities around the world to advance genetic research in legumes and other crops, with many research gaps still yet outstanding. Another major application of GMOs is the production of legume-derivatives, which broaden the consumption range of these crops beyond their natural forms. These include legume flour (used for making pastries, snacks, and spreads), dairy products (soymilk, fermented dairy blends and infant formula), industrial productions (biodegradable plastics, oil, gums, dyes, and inks), food stabilisers, fabric thickeners *etc*.

However, to avoid setbacks, GM foods labelling was introduced by researchers in response to consumers concerns towards and give awareness so that consumers can choose to consume GM, non-GM foods or both [54]. This regulation is mandatory and labelling of GM foods became a policy to ensure food safety and give consumers a right to choose [55]. In China, the regulation on safety administration of agricultural GMOs ensured that every product that enters their country contain labels and have a safety certificate attached to it from the Chinese agricultural ministry [56]. Some countries like India have put in place biosafety regulatory systems ensuring that thorough research on GM crops before commercialisation take place. These included the establishment of several government and non-governmental laboratories that willingly invest their time on research and development of protocols on how to handle all transgenic crops [49].

Rigorous field-testing of GM crops in confined and isolated areas is mandatory as part of the regulations which have been introduced as a way to affirm people that their concerns of safety are considered. The European Union has also introduced GMO regulations for detection, requiring farmers to disclose their methods of GMO detection for validation and verification by relevant authorities before consumption [57]. One of the common regulations used by most countries is that all GMO products that enters each country should be authorised before entry. According to Prakash *et al.* [58], the Cartagena Protocol on Biosafety ensures that all potential risks associated with GMOs go through thorough assessment and well-maintained transparency. The Cartagena Protocol on Biosafety fulfil their purpose of risk assessment by requesting detailed information on the GMO product used. This include molecular and genetic characteristic properties of genomic products used and their potential effects, nutritional value, allergenicity as well as toxicity effects [58].

PERCEPTIONS AND CONTROVERSIES SURROUNDING THE USE OF GMO

Amidst the advantages lies several concerns and possible shortcomings that come with GM crop production at farm-level. Garcia-Yi *et al.* [10] pointed out that the common practice of exchanging seeds between farmers get disrupted by the purchase of GM seeds, which can interfere with the rapport between local and international farmers or co-farmers. Some reports, furthermore, highlighted that pesticide resistant cultivars induce allergic reactions due to possible chemical residues found in GM foods. As such, in countries where plant biotechnology is scrutinised, ethical concerns (*i.e.* intellectual property rights, restrictions and regulations) limit farmers in the cultivation of GM crops and this can be a major setback for potential GM beneficiaries. The continued ban of GM crop commercialisation will have negative impacts on consumers. These restrictions drive the cost of GM crops and GM containing foods, to a slightly higher price than that of the wild type legumes. These affects people in different ways depending on whether they can afford and their household income.

According to Brookes and Barfoot [6], the adoption of pesticide resistant GM crops in 1996 led to a reduction in CO_2 emissions due to reduced fuel use and additional soil carbon sequestration. However, greater efforts and benefits of using pesticide resistant crops will be fully realised, following good human behaviour, and thorough research of their impacts on sustainable biological, ecological, and chemical interactions. Thus, it remains essential for people to know what they cultivate and consume, particularly in terms of the composition and the manufacturing process of the product. Consumers have to be made aware when new strategies are introduced in GM food production industries, to know their compositions and possible side effects, if any [59]. As consumers understands that, GM foods contain modified genes, some raise many opinions and false alarm about their compositions and processing. Many countries have therefore, conducted research on consumer perceptions of GM foods, with the outcomes indicating that people are either ignorant, fearful or have unjustified concerns about the GMOs.

In South Africa, research conducted on urban consumers' attitude towards GM white maize indicated that, majority of people were positive about GM white maize while fewer people were concerned [60]. The positive consumer perception in developing countries is largely owed to the transfer of knowledge, emphasising that this technology can address the persistent challenge of malnutrition and undernourishment. Consumers in the European Union and some parts of Asia mostly have negative attitudes towards GM foods due to the unknown health and environmental consequences imposed by some government entities and non-governmental organisations [61]. However, some speculative reports suggest that this kind of consumer attitude is fuelled only by prices of GM foods over non-GMOs and battles in agribusiness, especially organic farmers, and GM crop growers. Although, a recent survey done in Europe indicated that over 50% consumers are still in opposition of GM foods. Many consumers around the world and in the United States are still more positive about GM foods, in addition to organic foods [62, 63].

Most consumers were concerned about the GM food safety despite having the positive attitudes towards the biotechnological techniques used [56]. Furthermore, GM food labelling has been one of the concerns, and consumers suggested that food labelling should be made mandatory because people have the right to know if they are consuming GM food or not [55]. The various surveys conducted also indicated that consumers are against GM foods because of increased ignorance to the benefits in agricultural production and the lack-of or little scientific knowledge about GMOs, as well as the techniques used in their production. Above all, the major concern lies with transgenic crops, which may involve the use of plasmid vectors, such as in *Agrobacterium*-mediated genetic transformation protocols, whose sources are mainly bacteria and viruses. These microorganisms are typically associated with the stigma of pathogenicity, parasitism, and mortality, even though some are proven to be beneficial to human nutrition and metabolism. As indicated by Gastrow *et al.* [64], issues of genetic integrity, biosafety, moral beliefs, and unanticipated mutations, believed to be emanating from the consumption of GM foods, are among the controversial issues faced by biotechnology. According to Lusk and Coble [65] and Qaim and de Janvry [66], the main concern expressed by members of the public, scientists and medical professionals is the issue of biosafety. Other suggested risks include environmental harm, disruption of the natural agricultural practices, moral concerns, and unwarranted dominance in the commercial sector.

CONCLUSION AND FUTURE PERSPECTIVE

The leguminous grain crops have continued to play a significant role in the daily diets of many people worldwide, as a result of their high amount of nutritional primary and secondary metabolites. Legumes serve as the second largest family (*Fabaceae*) of flowering crop plants cultivated globally and explored as model crops for scientific studies. Legume crops are prepared as healthy food meals in Indian, Latin American, African, Mediterranean, and Middle Eastern cuisines, including dairy and fermented food products. The industries use them in flour milling to manufacture confectionery, soups, pastries, fodder, and remedy ingredient for diarrhea and dysentery. These widespread uses of legumes are primary indicators of their cost-effective and feasible production processes, accompanied by the ease of cultivation, growth maintenance, and yield harvesting for both smallholder and commercial farmers [1, 4].

Crop biotechnology has advanced over the years as new knowledge of legume plant biological systems and their functions has unraveled. Their benefits to human health, farmers, and biotechnologists, the economy, and mostly food security encourage novel developments to enhance the genetic and genomic properties of the crops, leading to naturally rich sources of essential phytonutrients and continued use as staple foods in various developing countries. To maximise these benefits, research has to be expanded with the aim of improving the nutritional content of plants, confer disease and abiotic stress resistance, increase seed storage viability, herbicides resistance, and antibiotics tolerance [2, 6, 10]. In addition to these, and vital to food security approaches, there is an increasing demand for biofuel production around the world. This requires an improvement in the quantity of oil produced, which can be addressed by improving the quality of seed oils in legume crops, as suggested by Song *et al.* [53].

However, a large number of genotypes still remain recalcitrant to genetic manipulations, pending a cost-effective, precise, highly-competent, and robust approach for the generation of fertile transgenic plants. Molecular, agronomic, and conventional breeding techniques have played a dynamic role in the genetic transformation of value-added crop legumes, especially soybean, lentils, peas, chickpeas, kidney beans, common beans, black gram, black-eyed peas, and pinto beans, amongst others. Major legume species (soybean, alfalfa, peanut, and pea) have undergone multiple trial tests for commercialisation and assessing the effects of GM crops on human health and the environment, some even dating back to 1996 [43]. These tests provide the scientific evidence needed to support and encourage the continued ban or use in the lifting of restrictions on the use of genetically modified organisms.

In conclusion, it is crucial that plant biotechnologists devise stable and functional genetic transformation protocols for all major legumes considering their widespread use as staple foods by people in developing countries. Moreover, the successful engineering of grain legumes will not only benefit these countries at the household level but at the economic level as they stand to play a major critical role in addressing the negative impacts of poverty, unemployment, hunger, diseases, food insecurity, and inequality.

LIST OF ABBREVIATIONS

FAO	Food and Agriculture Organisation
GMOs	Genetically modified organisms
GUS	Glucuronidase
ICRISAT	International crops research institute for the semi-arid tropics
ISAAA	International service for the acquisition of agri-biotech applications
Mrna	Messenger ribonucleic acid
NPTII	Neomycin phosphotransferase
PEG	Polyethylene glycol
PEM	Protein-energy malnutrition
RNAi	Ribonucleic acid interference
ROS	Reactive oxygen species

CONSENT FOR PUBLICATION

Not applicable.

CONFLICT OF INTEREST

The author declares no conflict of interest, financial or otherwise.

ACKNOWLEDGEMENTS

Declared none.

REFERENCES

[1] Tran LSP, Nguyen HT. Future biotechnology of legumes.Nitrogen fixation in crop production. Madison, WI, USA: Am Soc Agronomy, Crop Sci Soc Am, Soil Sci Soc Am 2009; pp. 265-308.

[2] Christou P. The biotechnology of crop legumes. Euphytica 1994; 74: 165-85. [http://dx.doi.org/10.1007/BF00040399]

[3] Shea Z, Singer SW, Zhang B. Soybean production, versitality and improvement.Legume crops. London: IntechOpen 2019; pp. 1-23.

[4] Maphosa Y, Jideani VA. The role of legumes in human nutrition.Functional food- improve health through adequate food. London: IntechOpen 2017; pp. 104-21. [http://dx.doi.org/10.5772/intechopen.69127]

[5] Sutivisedsak N, Moser BR, Sharma BK, *et al.* Physical properties and fatty acid profiles of oils from black, kidney, great-northern and pinto beans. J Am Oil Chem Soc 2010; 88: 193-200. [http://dx.doi.org/10.1007/s11746-010-1669-8]

[6] Brookes G, Barfoot P. Economic impact of GM crops: the global income and production effects 1996-2012. GM Crops Food 2014; 5(1): 65-75. [http://dx.doi.org/10.4161/gmcr.28098] [PMID: 24637520]

[7] Somers DA, Samac DA, Olhoft PM. Recent advances in legume transformation. Plant Physiol 2003; 131(3): 892-9. [http://dx.doi.org/10.1104/pp.102.017681] [PMID: 12644642]

[8] International service for the acquisition of agri-biotech applications (ISAAA). Global status of commercialised biotech/GM crops, ISAA Brief No 54. United States, Ithaca, New York 2018.

[9] Food and agriculture organisation of the united nations (FAO). GMOs and the food supply chain. Rome: Report of the Codex Alimentarius commission 2000.

[10] Garcia-Yi J, Lapikanonth T, Vionita H, *et al.* What are the socio-economic impacts of genetically modified crops worldwide? A systematic map protocol. Environ Evid 2014; 3(24): 1-17. [http://dx.doi.org/10.1186/2047-2382-3-24]

[11] Raman R. The impact of genetically modified (GM) crops in modern agriculture: A review. GM Crops Food 2017; 8(40): 195-208.

[12] Azadi H, Samiee A, Mahmoudi H, *et al.* Genetically modified crops and small-scale farmers: Main opportunities and challenges. Crit Rev Biotechnol 2015; 1-13. [http://dx.doi.org/10.3109/07388551.2014.990413] [PMID: 25566797]

[13] Polak R, Phillips EM, Campbell A. Legumes: Health benefits and culinary approaches to increase intake. Clin Diab J 2015; 33(4): 198-205.

[14] Sanderson LA, Caron CT, Tan R, Shen Y, Liu R, Bett KE. Know pulse: a web-resource focused on diversity data for pulse crop improvement. Front Plant Sci 2019; 10(965): 965. [http://dx.doi.org/10.3389/fpls.2019.00965] [PMID: 31428111]

[15] Pathak N. Health benefits of legumes. Accessed date 27 march 2020. Available online at https://www.webmd.com/foo-recipes/health-benefits-legumes

[16] Mabaleha MB, Yeboah S. Characterization and compositional studies of the oils from some legume cultivars, *Phaseolus vulgaris*, grown in Southern Africa. J Am Oil Chem Soc 2004; 81: 361-4. [http://dx.doi.org/10.1007/s11746-004-0907-6]

[17] Amanlou H, Maheri-Sis N, Bassiri S, *et al.* Nutritional value of raw soybeans, extruded soybeans,

roasted soybeans and tallow as fat sources in early lactating dairy cows. Open Vet J 2012; 2(1): 88-94. [PMID: 26623299]

[18]   Wieczorek AM, Wright MG. History of agricultural biotechnology: How crop development has evolved. Nat Edu Know 2012; 3(3): 1-9.

[19]   Finer J, Dhillon T. Transgenic plant production.Plant biotechnology and genetics: Principles, techniques and applications. John Wiley and Sons, Inc. 2008; pp. 245-73.
[http://dx.doi.org/10.1002/9780470282014.ch10]

[20]   Paz MM, Shou H, Guo Z, Zhang Z, Benerjee AK, Wang K. Assessment of conditions affecting *Agrobacterium*-mediated soybean transformation using the cotyledonary node explant. Euphytica 2004; 136: 167-79.
[http://dx.doi.org/10.1023/B:EUPH.0000030670.36730.a4]

[21]   Zia M, Arshad W, Bibi Y, Nisa S, Chaudhary MF. Does *Agro*-injection to soybean pods transform embryos? Omics J 2011; 4(7): 384-90.

[22]   Zhang X, Henriques R, Lin SS, Niu QW, Chua NH. Agrobacterium-mediated transformation of Arabidopsis thaliana using the floral dip method. Nat Protoc 2006; 1(2): 641-6.
[http://dx.doi.org/10.1038/nprot.2006.97] [PMID: 17406292]

[23]   Cheng M, Fry JE, Pang S, *et al.* Genetic transformation of wheat mediated by *Agrobacterium tumefaciens.* Plant Physiol 1997; 115(3): 971-80.
[http://dx.doi.org/10.1104/pp.115.3.971] [PMID: 12223854]

[24]   Hinchee MAW, Connor-Ward DV, Newell CA, *et al.* Production of transgenic soybean plants using *Agrobacterium*-mediated DNA transfer. Nat Biotechnol 1988; 6: 915-22.
[http://dx.doi.org/10.1038/nbt0888-915]

[25]   Zeng P, Vadnais DA, Zhang Z, Polacco JC. Refined glufosinate selection in Agrobacterium-mediated transformation of soybean [*Glycine max* (L.) Merrill]. Plant Cell Rep 2004; 22(7): 478-82. [*Glycine max* (L.) Merill.].
[http://dx.doi.org/10.1007/s00299-003-0712-8] [PMID: 15034747]

[26]   Yi X, Yu D. Transformation of multiple soybean cultivars by infecting cotyledonary-node with *Agrobacterium tumefaciens.* Afr J Biotechnol 2006; 5(20): 1989-93.

[27]   Liu SJ, Wei ZM, Huang JQ. The effect of co-cultivation and selection parameters on *Agrobacterium*-mediated transformation of Chinese soybean varieties. Plant Cell Rep 2008; 27(3): 489-98.
[http://dx.doi.org/10.1007/s00299-007-0475-8] [PMID: 18004571]

[28]   Untergasser A, Bijl GJM, Liu W, Bisseling T, Schaart JG, Geurts R. One-step Agrobacterium mediated transformation of eight genes essential for rhizobium symbiotic signaling using the novel binary vector system pHUGE. PLoS One 2012; 7(10): e47885.
[http://dx.doi.org/10.1371/journal.pone.0047885] [PMID: 23112864]

[29]   Tripathi L, Singh AK, Singh S, *et al.* Optimization of regeneration and *Agrobacterium*-mediated transformation of immature cotyledons of chickpea (*Cicer arietinum* L.). Plant Cell Tissue Organ Cult 2013; 113: 513-27.
[http://dx.doi.org/10.1007/s11240-013-0293-3]

[30]   Ko TS, Nelson RL, Korban S. Screening multiple soybean cultivars (MG00 to MG VIII) for somatic embryogenesis following *Agrobacterium*-mediated transformation of immature cotyledons. Crop Sci 2004; 44: 1825-31.
[http://dx.doi.org/10.2135/cropsci2004.1825]

[31]   Olhoft PM, Somers DA. L-Cysteine increase *Agrobacterium*-mediated T-DNA delivery into soybean cotyledonary-node cells. Plant Cell Rep 2001; 20: 706-11.
[http://dx.doi.org/10.1007/s002990100379]

[32]   Chowrira GM, Akella V, Fuerst PE, Lurquin PF. Transgenic grain legumes obtained by *in planta* electroporation-mediated gene transfer. Mol Biotechnol 1996; 5(2): 85-96.

[http://dx.doi.org/10.1007/BF02789058] [PMID: 8734422]

[33]   Rivera AL, Gómez-Lim M, Fernández F, Loske AM. Physical methods for genetic plant transformation. Phys Life Rev 2012; 9(3): 308-45.
[http://dx.doi.org/10.1016/j.plrev.2012.06.002] [PMID: 22704230]

[34]   Quecini VM, de Oliveira CA, Alves AC, Vieira MLC. Factors influencing electroporation-mediated gene transfer to *Stylosanthes guianensis* (Aubl.) Sw. protoplasts. Genet Mol Biol 2002; 25(1): 73-80.
[http://dx.doi.org/10.1590/S1415-47572002000100014]

[35]   Sanford JC. The biolistic process. Trends Biotechnol 1988; 6: 299-302.
[http://dx.doi.org/10.1016/0167-7799(88)90023-6]

[36]   Boynton JE, Gillham NW, Harris EH, *et al.* Chloroplast transformation in *Chlamydomonas* with high velocity microprojectiles. Science 1988; 240(4858): 1534-8.
[http://dx.doi.org/10.1126/science.2897716] [PMID: 2897716]

[37]   Bhargava SC, Smigocki AC. Transformation of tropical grain legumes using particle bombardment. Curr Sci 1994; 66(6): 439-42.

[38]   Aragão FJL, Barros LM, Brasileiro AC, *et al.* Inheritance of foreign genes in transgenic bean (*Phaseolus vulgaris* L.) co-transformed *via* particle bombardment. Theor Appl Genet 1996; 93(1-2): 142-50.
[http://dx.doi.org/10.1007/BF00225739] [PMID: 24162211]

[39]   Hefferon KL. Can biofortified crops help attain food security? Curr Mol Biol Rep 2016; 1(4): 180-5.
[http://dx.doi.org/10.1007/s40610-016-0048-0]

[40]   McGloughlin MN. Modifying agricultural crops for improved nutrition. N Biotechnol 2010; 27(5): 494-504.
[http://dx.doi.org/10.1016/j.nbt.2010.07.013] [PMID: 20654747]

[41]   World Health Organisation (WHO). Food, genetically modified. Date Accessed: 31 march 2020. Available Online at http:www.who.int/topics/food_genetically_modified/en/

[42]   Abady S, Shimelis H, Janila P. Farmers' perceived constraints to groundnut production, their variety choice and preferred traits in eastern Ethiopia: Implications for drought-tolerance breeding. J Crop Improv 2019; 33(4): 505-21.
[http://dx.doi.org/10.1080/15427528.2019.1625836]

[43]   Dunwell JM. Transgenic approaches to crop improvement. J Exp Bot 2000; 51(Spec No): 487-96.
[http://dx.doi.org/10.1093/jexbot/51.suppl_1.487] [PMID: 10938856]

[44]   Buiatti M, Christou P, Pastore G. The application of GMOs in agriculture and in food production for a better nutrition: two different scientific points of view. Genes Nutr 2013; 8(3): 255-70.
[http://dx.doi.org/10.1007/s12263-012-0316-4] [PMID: 23076994]

[45]   Petrovic G, Nikolic Z. Genetically modified crops- A potential risk for sustainable agriculture. Chapter 2013; pp. 289-97.

[46]   Zdjelar G, Nikolić Z, Vasiljević I, *et al.* Detection of genetically modified soya, maize, and rice in vegetarian and healthy food products in Serbia. Czech J Food Sci 2013; 31: 43-8.
[http://dx.doi.org/10.17221/105/2012-CJFS]

[47]   De Ron AM, Cubero JI, Singh SP, Aguilar OM. Cultivated legume species. Int J Agron 2013; 324619.

[48]   Mureithi JG, Gachene CKK, Ojiem J. The role of green manure legumes in smallholder farming systems in Kenya: The legume research network project. Trop Subtrop Agroecosystems 2003; 1: 57-70.

[49]   Smyth SJ. Genetically modified crops, regulatory delays, and international trade. Food Energy Secur 2016; 6(2): 78-86.
[http://dx.doi.org/10.1002/fes3.100]

[50] Shukla M. AL-Busaidi KT, Trivedi M, Tiwari RK. Status of research, regulation and challenges for genetically modified crops in India. GM Crops Food 2018; 9(4): 173-88.
[http://dx.doi.org/10.1080/21645698.2018.1529518] [PMID: 30346874]

[51] Forster D, Andres C, Verma R, Zundel C, Messmer MM, Mäder P. Yield and economic performance of organic and conventional cotton-based farming systems--results from a field trial in India. PLoS One 2013; 8(12): e81039.
[http://dx.doi.org/10.1371/journal.pone.0081039] [PMID: 24324659]

[52] Dita MA, Rispail N, Prats E, Rubiales D, Singh KB. Biotechnology approaches to overcome biotic and abiotic stress constraints in legumes. Euphytica 2006; 147: 1-24.
[http://dx.doi.org/10.1007/s10681-006-6156-9]

[53] Ochatt SJ. Agroecological impact of an *in vitro* biotechnology approach of embryo development and seed filling in legumes. Agron Sus Dev J 2015; 35: 535-52.
[http://dx.doi.org/10.1007/s13593-014-0276-8]

[54] Song Y, Wang XD, Rose RJ. Oil body biogenesis and biotechnology in legume seeds. Plant Cell Rep 2017; 36(10): 1519-32.
[http://dx.doi.org/10.1007/s00299-017-2201-5] [PMID: 28866824]

[55] Viljoen CD, Dajee BK, Botha GM. Detection of GMO in food products in South Africa: Implications of GMO labelling. Afr J Biotechnol 2006; 5(2): 73-82.

[56] Rosculete E, Bonciu E, Rosculete CA, Teleanu E. Detection and quantification of genetically modified soybean in some food and feed products. A case study on products available on Romanian market. Sustainability 2018; 10(1325): 1-13.
[http://dx.doi.org/10.3390/su10051325]

[57] Li Q, Curtis KR, McCluskey JJ, Wahl TI. Consumers attitudes towards genetically modified foods in Beijing, China. AgBioForum 2002; 5(4): 145-52.

[58] Davison J, Ammann K. New GMO regulations for old: Determining a new future for EU crop biotechnology. GM Crops Food 2017; 8(1): 13-34.
[http://dx.doi.org/10.1080/21645698.2017.1289305] [PMID: 28278120]

[59] Prakash D, Verma S, Bhatia R, Tiwary BN. Risks and precautions of genetically modified organisms. ISRN Ecol 2017; 1-13.
[PMID: 369573]

[60] Wunderlich S, Gatto KA. Consumer perception of genetically modified organisms and sources of information. Adv Nutr 2015; 6(6): 842-51.
[http://dx.doi.org/10.3945/an.115.008870] [PMID: 26567205]

[61] Vermeulen H, Kirsten JF, Doyer TO, Schonfeldt HC. Attitudes and acceptance of South African urban consumers towards genetically modified white maize. Agrekon 2015; 44(1): 118-37.
[http://dx.doi.org/10.1080/03031853.2005.9523705]

[62] Curtis KR, McCluskey JJ, Wahl TI. Consumer acceptance of genetically modified food products in the developing world. AgBioForum 2004; 7: 70-5.

[63] Deng H, Hu R, Pray C, Jin Y. Perception and attitude toward GM technology among Agribusiness managers in China as producers and as consumers. Sustainability MDPI J 2019; 11(5): 1-17.
[http://dx.doi.org/10.3390/su11051342]

[64] Gastrow M, Roberts B, Reddy V, Ismail S. Public perceptions of biotechnology in South Africa. S Afr J Sci 2018; 114(1/2): 1-9.
[http://dx.doi.org/10.17159/sajs.2018/20170276]

[65] Lusk JL, Coble KH. Risk perceptions, risk preference, and acceptance of risky food. Amer J Agri Econ. Agric Appl Econ Ass 2005; 87(2): 393-405.
[http://dx.doi.org/10.1111/j.1467-8276.2005.00730.x]

[66] Qaim M, de Janvry A. Genetically modified crops, corporate pricing strategies, and farmers' adoption: The case of Bt cotton in Argentina. Amer J Agri Econ, Agric Appl. Econ Ass 2003; 85(4): 814-28.

SUBJECT INDEX

A